ROUNDWOOD
TIMBER FRAMING
Building Naturally Using Local Resources

Ben Law

Published by
Permanent Publications
Hyden House Ltd
The Sustainability Centre
East Meon
Hampshire GU32 1HR
United Kingdom
Tel: 01730 776 582
Overseas: (international code +44 - 1730)
Email: enquiries@permaculture.co.uk
Web: www.permaculture.co.uk

Published in association with The Sustainability Centre
www.sustainability-centre.org

Distributed in North America by
Chelsea Green Publishing Company, PO Box 428, White River Junction, VT 05001
www.chelseagreen.com

© 2010 Ben Law
Reprinted 2016, 2019
The right of Ben Law to be identified as the author of this work has been asserted by him in accordance with the Copyrights, Designs and Patents Act 1998

Designed and typeset by Two Plus George Limited, www.TwoPlusGeorge.co.uk
Cover design by John Adams

Printed in the UK by CPI Antony Rowe, Chippenham, Wiltshire

All paper from FSC certified mixed sources

The Forest Stewardship Council (FSC) is a non-profit international organisation established to promote the responsible management of the world's forests. Products carrying the FSC label are independently certified to assure consumers that they come from forests that are managed to meet the social, economic and ecological needs of present and future generations.

British Library Cataloguing-in-Publication Data
A catalogue record for this book is available from the British Library

ISBN 978 1 85623 330 9

All rights reserved. No part of this publication may be reproduced, stored in a retrieval system, rebound or transmitted in any form or by any means, electronic, mechanical, photocopying, recording or otherwise, without the prior permission of Hyden House Limited.

This project has received grant aid from The European Agricultural Development Fund: Europe investing in rural areas. The Fieldfare Local Action Group manages RDPE LEADER funding across the rural areas of East Hampshire & Winchester.
www.fieldfareleader.org.uk

Contents

Foreword by Lloyd Kahn	iv
Introduction	vi
1 Roundwood in Building	1
2 Managing Woodlands for Roundwood Timber Framing	9
3 Tree Species for Roundwood Timber Framing	17
4 Tools for Roundwood Timber Framing	37
5 Construction	59
6 Beyond the Frame	93
7 Roundwood Timber Framing Builds	111
Extroduction	146

Appendices

A	Lodsworth Larder: Extracts from the Engineering Report	149
B	Acknowledgements	156
C	Glossary	156
D	Further Reading	158
E	Resources	159

Foreword

BY LLOYD KAHN

MY FIRST EXPERIENCE with house building was in 1960. What began as a simple remodeling job on a summer cottage in California ended up leading me into almost 50 years of recording owner-builders and indigenous buildings in many parts of the world.

Disappointed with ego-driven architects and their grand-scale homes, I've been drawn to buildings sited and built with local materials, created by owner-builders and designed with practicality, aesthetics and ecology in mind — homes created with one's own hands.

Homes that, not incidentally, feel good to be inside and where the human spirit can thrive.

* * *

About six months ago, I ordered a copy of Ben Law's book, *The Woodland House*. When I saw a picture of his round wood pole home with waney edged board siding in the west Sussex woods, I got a jolt. Once in a while, a building will just stop me in my tracks, where everything works to create a harmonious whole – design, materials, craftsmanship, siting. Here was such a creation. Here was a kindred spirit.

Ben's work is unique. He understands the history of timber framing, starting with cruck frames, box frames and log houses. He's a 21st century woodsman, formulating a new building style based on the timber framing methods of his British ancestors, but incorporating new techniques and the realities of the present-day world. He's created a new vernacular, one that he's passing on to future builders via hands-on workshops and books like these.

Ben starts by sourcing wood right from the building site and, with every element in a construction, he seeks practical solutions using materials in their most natural form and that are the least harmful to the planet. There is also a sense of continuity in planting, tending and harvesting of timber for future projects and future generations.

In these suddenly green-conscious days, there's a revival of interest in owner-building and the use of natural materials. The timing of this exciting book is perfect. Thumb through the pages and you'll see the creation of natural buildings – from the selection of locally grown timber, to harvesting, construction of round pole structural timber frames and the finished buildings themselves. This is a formula that can be used for building your own home with your own hands, step-by-step, and in connection with the rhythms of the planet.

Lloyd Kahn
July 2010

Lloyd Kahn is the editor of Shelter *(1973),* HomeWork: Handmade Shelter *(2002), and* Builders of the Pacific Coast *(2008). His latest book on building is* The Gardeners' and Poultry Keepers' Guide, *a reprint of a 10 year-old London catalogue of greenhouses, chicken coops, and small cottages. He's now working on a book on tiny houses.*

Introduction

The building industry is one of the most high energy, high waste industries in the UK. Although building regulations are helping to create better insulated buildings, many of the components are mass produced and incur high building miles, being transported vast distances to the construction site. Mass produced fixings and features often produce soulless architecture and the social atmosphere of many building sites is far from welcoming. The building industry ploughs forward in a reactionary fashion, making small changes and concessions through new regulations but not stopping to consider methods that are low impact, locally sourced and workforce friendly.

Roundwood timber framing is one such technique. Working with materials sourced straight from the forest, using nature's shapes to create the structure, using techniques that can easily be learnt and passed on and a method of building that can continue without stalling in the transition to a post oil society.

The need to express one's creativity through freeform building and a need to comply to plans and regulations has always been a difficult balance, but through the evolution of roundwood timber framing, I believe I am finding a style and technique which satisfies both needs. Hand selecting trees with form and character that have their own intrinsic beauty and follow their own lines, rather than those that have been forced upon them by the saw and right angle, allows freedom of movement in a building whilst keeping within the parameters of the drawings on the table. The building itself has life, curves and natural form, the frames often looking like they are trees growing out of the floorboards. Each new building improves on the last and each joint is developed and refined. I feel roundwood timber framing has reached a point in its

Five cruck roundwood frame with large verandah under construction.

evolution where the joints are advanced, the timbers tried and tested and a range of buildings including sheds, barns, dwellings, educational spaces and industrial buildings have been constructed and passed the vigorous analysis of the construction engineers and building inspectors.

So it is time to pass on the fruits of this work, encourage you to search your local woodlands for timber, and create roundwood timber framed buildings in our landscape, which leave a legacy of hope to future generations. Roundwood timber framing is in its infancy, but what you will find here has its roots firmly grounded in the woodsman traditions of England.* It is the love and knowledge of woodsmanship and its revival that has spurred us woodsmen to come out of the wood for periods during the spring and summer, to build our roundwood creations before returning to the woodland landscape to cut the coppice and continue the renaissance of the British forest dweller.

* 'Woodsman' is not a reference to gender, as men and women are equally capable of woodland activities. It comes from *manus*, the Latin word meaning 'hand'. Thus woodsman means 'hand of the woods'.

Chapter 1

Roundwood in Building

THE FIRST BUILDING that I constructed and lived in was a bender; bent coppiced hazel rods lashed together with string and then covered with a tarpaulin. It was a simple structure that more than sufficed as a home in the woods for a few years.

My next home was a yurt; a much more intricate framework of poles, advanced further by the use of steam to shape and bend the poles to extremes I could never have managed with the more simple bending of greenwood poles that formed my bender.

My next timber home was the 'Woodland House'; a roundwood timber frame structure. I had evolved through the process of building to create a comfortable and aesthetically pleasing home, without straying from the use of roundwood.

Was it just that I was so well immersed with life in the woods, surrounded by trees and their natural form, that I missed the step where humans moved on to build with sawn wood? Or perhaps it is through working with roundwood, that I have found a building practice that not only benefits woodsmen and foresters, but lays down a benchmark for a new architectural vernacular in low impact sustainable building?

I believe it is the latter. I have constructed sawn timber frames from green oak to gain ideas for the traditional framing joints. I have experimented with the scribe joints of the cabin builders of Canada and the United States of America. Both techniques are tried and tested over many years to produce solid and long lasting buildings but both have their limitations when we look to the future, at our available resources and the need to build from what is readily available around us.

TOP LEFT Author's yurt.

BOTTOM LEFT Author's first home at Prickly Nut Wood — a 'bender'.

MAIN Author outside the Woodland House.

ROUNDWOOD TIMBER FRAMING 3

TIMBER FRAMING

Britain has a wonderful ancestry of timber framing tradition. Early timber buildings consisted of vertical posts placed upright in a trench which was then back-filled with earth. The effort of hauling timber over long distances soon helped to evolve the buildings by spacing the vertical log poles further apart and infilling with twigs and clay, an early form of wattle and daub. Attaching the roof to such a structure was more difficult and it became necessary to attach a wall plate across the top of the posts upon which the roof rafters could then be attached. With evolution it became clear that the posts no longer needed to be earthed into a trench but could be attached to a sill place below them to keep the building stable and avoid the log posts rotting in the earthen trench.

So the timber frame was formed and over time the skill and knowledge of medieval carpenters constructed some of the most beautiful and long lasting buildings in Britain.

The woodlands of Britain were the source for these buildings and coppice woodlands, with their productive yield of underwood for fires, charcoal and craft produce, aided the management of standard trees. These were predominately oak, which were felled and converted into the structural timbers for the traditional timber frame.

Early frames were often 'cruck' frames. These were formed from a naturally curved tree, split down its length with the two mirroring halves joined at the peak. Pairs of crucks were then joined together to form a primitive frame. The addition of a collar or tie beam formed the A shape which give the cruck frame its strength and stability. In Britain, there are over 3,000 examples of these frames still in use. With improved tools and jointing, carpenters were able to create frames without the need for the curved cruck and hence the box frame was formed.

The simplest box framed structures consisted

of corners and intermediate posts rising from the sill plate to the wall plate with tie beams running across at each end of the frame and at chosen intervals across the intermediate posts. Bracing was often achieved with wind braces, sometimes curved to avoid the frame from racking. One advantage of box frame construction was that the internal space could be squared or regularised as there were no protruding timbers as in cruck construction. A well designed building, however, can also make good aesthetic and practical use of crucks within the building.

The box frame has formed the basis for the nost of the timber frames built up to the present day, the majority of them being built from oak. Oak needs to be grown to a good age, 70 to 80 years in the UK, before it is good enough to use for timber framing. This is because of the large amount of sap wood that the tree forms during its early years of growth. Other species have been used in the past, and in the future it will be necessary to use species grown as locally as possible to the construction site, opening up the possibilities of mixed species frames.

LOG BUILDING IN CONSTRUCTION

To many, the image of the log cabin, simply constructed, holds a romantic rural impression of Scandinavia, North America and Canada. These simple structures, formed from the abundant forest, made early shelters and hunting retreats. This basic form is a simplistic version of the art of log building which produces some of the most beautiful log houses, where the skill and craftsmanship of log joining is a testament to the carpenter's knowledge.

Construction is by placing logs horizontally and building upwards, notching together each log to form strong and stable walls. Scribing tools are used to ensure accurate jointing of the logs. The corners are notched together to form a strong joint that avoids the need for other fastenings.

A log building for a small family takes at least a year to complete, more ambitious projects taking up to three years. The volume of logs (trees) used is substantial and although the thickness of the logs saves on insulation, it is likely to be deemed a decadent use of resources as we look towards the future.

OPPOSITE TOP Coppice with standards. Oak standard over regrowing chestnut coppice.

OPPOSITE BOTTOM Coppice with standards, freshly cut.

ABOVE LEFT Log cabin.

ABOVE RIGHT Traditional oak timber frame.

CARPENTER'S KNOWLEDGE

Carpenters have learnt their craft over many generations. They have evolved specific tools to ease and improve their craft and in more recent times have worked to plans rather than building freeform. A traditional carpenter would deal with a problem as it occurred, adjust the pitch or angle of a roof, evolve the joint of a beam, and the end result would be a handmade house, some looking much more handmade than others. With modern buildings, plans are drawn up, problems are anticipated and resolved and one building will look similar to the next one.

Traditional carpenters had a knowledge of timber and how it was grown. Today's modern carpenter knows what a tanilised softwood rafter of 150mm by 50mm looks like and how to fix it together with a steel plate, but is unlikely to know what species of tree the timber came from; what altitude it grew at and how that can affect the structural integrity of the timber. A traditional carpenter would choose his or her trees while they were standing in the forest. They would be felled during the winter, so that the timber could be worked, still green, at the start of the building season, which began in spring. The seasonal pattern of work flowed with the weather. The winter spent cutting, rolled into a spring of framing, a summer of roofing and an autumn to complete ready to return to the woods as the next winter approached, leaving the building for its new inhabitants before the onset of cold weather.

THE ROUNDWOOD FRAME

My interest in building the roundwood frame came from the availability of timber trees growing within my woods. Being a predominately coppiced woodland, there were not the volume of straight trees to construct a log building

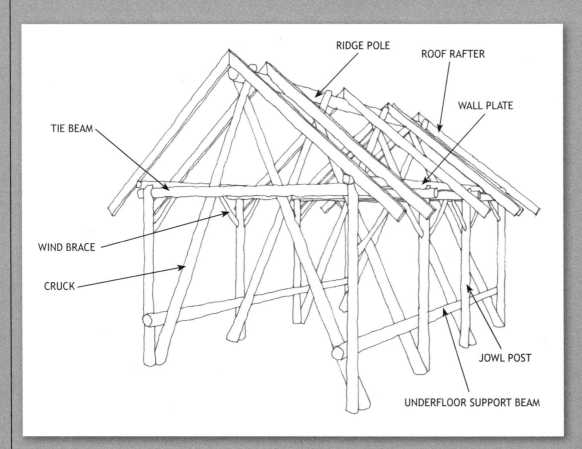

Line drawing showing key components of a roundwood timber frame.

Roundwood frame, Lodsworth Larder.

As I write this, I see a time not far in our future where the conventional building industry and transport haulage systems that move materials around the globe will have ground to a halt. We will need to prioritise our uses of the last remaining oil available and will have to rethink how we construct our homes and other buildings. The need to build from what is around us and the use of draught animals in haulage and hand tools in construction is an area of knowledge we need to refamiliarise ourselves with now.

STRENGTH OF ROUNDWOOD

A timber pole is stronger than sawn timber of equal cross sectional area, because fibres flow smoothly around natural defects and are not terminated as sloping grain at cut surfaces.

Lionel Jayenetti,
Timber Research and Development Agency

and neither was there an abundance of oak to produce a traditional timber frame.

The roundwood frame was therefore formed, taking its design from the traditional cruck of the British timber frame and using scribing techniques obtained from the log builders of Canada and North America. What roundwood timber framing offers is an aesthetic frame using far less timber than a log building and far younger trees than a traditional oak frame. It also opens up the opportunity to use a variety of timber species and mix them where appropriate within the frame design.

These buildings are designed for the 21st century. They make use of local timber resources, reduce the need to transport timber and building materials over long distances, leave behind the polluting processes of timber preservative and show that foundations can be constructed without the need for concrete. Above all, the construction process is not complicated and with a bit of practise, the opportunity to empower people to build their own homes from materials around them is an achievable goal.

The advantages of using roundwood come from its strength, which is greater than that of sawn wood. The opportunity to use smaller diameter poles with the same structural strength of larger sections of sawn timber opens up opportunities to use small diameter timber from the forest without conversion. By using roundwood the cost and wastage of conversion by sawing are removed. The ability of roundwood to regrow, especially when grown in a coppice system, ensures a continuous supply of renewable poles. The majority of oak cut

for traditional sawn timber framing dies and replanting is needed.

Some of the best research in strength of roundwood timber in bending, compression buckling and tension can be found in the report, *Round Small Diameter Timber For Construction* (see Appendix D). The research for this European Union project was carried out in Finland, the Netherlands, Austria and the UK and gives some useful figures demonstrating the different strengths of round poles in construction. It also looks at different techniques for grading roundwood poles for construction from x-raying timber through to a more simple visual grading formula. The need for a more simple grading system is clearly apparent, as at present each building my company constructs is over engineered (using larger poles than are necessary) in order to comfortably meet building regulations. Some engineer's figures on buildings constructed by the Roundwood Timber Framing Company Ltd can be found in Appendix A.

Two well qualified professionals in the strength of roundwood have stated that roundwood can be 50% stronger than sawn wood of the same cross-sectional area.

Other important factors to consider when choosing round poles for timber framing are the age of the poles in relation to their diameter. For example, where they have been grown and how they have been managed will have a major effect on the strength of the timber. With a softwood like larch, it is possible to find a 15 year old pole and a 30 year old pole both of 200mm diameter. Having grown faster to that size, the rings of the 15 year old pole will be further apart while the rings of the 30 year old pole will be closer together, making the older pole more suitable for roundwood timber framing.

Forest management therefore has a very important influence on the quality of poles for this form of timber framing.

SPECIES	fm	Fc,0	Em	Ec,0
Scots pine	53.8/9.3	28.4/4.9	13.1/4.1	12.2/2.8
Larch	78.3/12.7	41.4/3.7	13.6/2.5	-----
Douglas fir	52.0/9.9	33.0/ --	11.1/2.4	11.0/ --

Means and standard deviations of bending and compression values of tested round timber at test moisture content. Bending strengths are high, lower 5th percentile ranging from 35 to 60 N/mm^2, all the mean values exceeding $50N/mm^2$. There is strong evidence that larch has the best mechanical properties amongst the tested material.
Figures from *Round Small Diameter Timber for Construction.*
(See Appendix D.)

SPECIES	Sample size	Pmean (kg/m^3)	P05 (kg/m^3)	fm,mean (N/mm^2)	Fm,05 (N/mm^2) by rank	Em,mean (KN/m^2)
Scots pine	100	529	461	54	39	14.9
Larch	178	580	509	85	63	14.3
Douglas fir	180	442	367	52	37	11.1

Bending characteristics adjusted to 12% MC in accordance with EN 384. Size adjustment not used.
Figures from *Round Small Diameter Timber for Construction.*
(See Appendix D.)

SPECIES	Sample size	Fc,mean (N/mm^2)	Fc,05 (N/mm^2)
Scots Pine	250	30.2	24.0
Larch	58	45.0	38.0
Douglas Fir	190	33.0	26.0

Compression characteristics adjusted to 12% MC in accordance with EN 384. Size adjustment not used.
Figures from *Round Small Diameter Timber for Construction.* (See Appendix D.)

Chapter 2

Managing Woodlands for Roundwood Timber Framing

Coppice

Coppice is the term used to describe the successional cutting of broadleaf woodland during the dormant winter period. In spring, when the sap rises, the stump (known as the stool) sends up new shoots which are grown

on for a number of years until they reach the desired size. They are then cut again during winter and the process repeats itself. The wood cut for the coppice is known as underwood and has for centuries supplied a variety of traditional products and supported a large rural workforce, from the cutter to coppice merchant; craftsman to purchaser. Coppice is a valuable crop and managed well, it can sustain more people per acre than any of the modern forestry alternatives. It is also a sustainable pattern of management, rarely needing any replanting, so the soil is not disturbed and therefore not subject to the risk of erosion. Nutrients are returned to feed the growth of the tree mainly through the annual leaf fall. Coppice creates a cyclical habitat and unique ecosystem, and is one of the few patterns of symbiosis in nature where humans are an important part of the relationship. In a well managed coppice, the stools are closely spaced, from about 4-6 feet (1.2-1.8m) apart and the ground is fully shaded by the leaves and coppice shoots. When this is cut, sunlight pours in, dormant seeds waiting for light emerge and different birds, animals and insectlife move into the newly created habitat.

The most suitable species for roundwood timber framing found growing within a coppice

LEFT Sweet chestnut swept butt found growing in derelict coppice.

MAIN On cycle sweet chestnut coppice at Prickly Nut Wood. The coppice in the foreground shows one year of re-growth, the coppice behind 14 years.

system is sweet chestnut. Sweet chestnut is a fast growing durable hardwood with a very small amount of sap wood and it has been used in construction as a 'poor man's' oak for many centuries. Finding good quality straight sweet chestnut can be difficult as the stems are not as uniform as timber grown in a plantation. Finding poles of 10m plus for ridge poles and wall plates is rare and I often use softwood timber (larch or Douglas fir) for these longer lengths and sweet chestnut for the crucks and tie beams. As with all forestry, straight stems are formed by a good stocking rate and the competition for light draws the stems upward. There are occasions where a poor stocking rate can be useful. If the stocking rate is poor, swept butts are formed and these can be useful when selecting crucks. The sweep in the base can be used to create a curve that gives character which is unavailable in straighter timber. Most of the sweet chestnut I use in roundwood timber framing is between 25 and 40 years of age. Some of this coppice is what we refer to as overstood coppice and therefore needs recoppicing to improve its yield and get it back onto rotation. Sourcing higher value poles for roundwood timber framing can help with this process. As I write, sitting in my sweet chestnut roundwood timber framed house,

This sweet chestnut log shows clear definition between the dark coloured heartwood and lighter sapwood. The thin creamy coloured line around the edge is the sapwood.

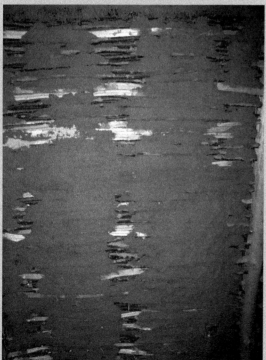

the coppice that was used in the frame is now nearly 10 years of age in regrowth. The stems are already 20 feet tall and 4 inches in diameter, and so the next house is growing as I write.

Ash is another species that I use internally in roundwood timber framing. It is a timber of great strength and works well for internal wind braces.

Hazel, one of the most ancient of all coppice trees, finds its place amongst the woven handrails and wattle and daub in the roundwood timber frame house.

When you approach construction with timber as I have, from the viewpoint of the forester, you already know so many of the strengths and characteristics of individual tree species and their way of growing. This makes choosing the right timbers for the right use a natural process. In the construction of the community village shop at Lodsworth, we used seven different timber species. All of these were sourced locally and chosen for their particular strengths and characteristics to fulfil a chosen role.

OPPOSITE Coppiced hazel visible both in the woven handrail infill and above in the gable end. The roof rafters are coppiced sweet chestnut.

ABOVE LEFT An internal windbrace formed from coppiced ash.

ABOVE RIGHT Coppiced hazel panels formed as part of an internal wall.

BELOW LEFT Earth plaster being applied to woven hazel internal wall panels.

Timber species	Uses in Lodsworth Larder
Sweet chestnut	Poles for main frame — crucks, jowels, tie beams, roundwood rafters, external wind braces, lathes
European larch	Wall plates, ridge pole, stud work
Douglas fir	Floor joists, roof rafters, stud work, stairs
Western red cedar	Shingles
Oak	Underfloor tie beam, pegs, floor boards, cladding
Ash	Internal wind braces
Hazel	Wattle infill

ROUNDWOOD TIMBER FRAMING 13

Plantations

Softwood plantations can offer poles that are ideal for roundwood timber framing. Most plantations are planted with the end goal of high value timber and have a number of thinnings throughout their lifetimes, the idea being that conifers which are closely planted will draw each other up to form straight poles with minimal side branches. A first thinning produces a crop of small diameter poles and allows the main plantation to increase in base diameter as well as height. A second thinning some years later further increases the base diameter. The exact timings of the thinnings will depend on species, altitude, soil type and other environmental factors. In most commercial operations, the first and sometimes second thinnings are an economic loss. The value of the poles thinned does not meet the cost of labour to thin the plantation and then to sell the poles at roadside or transport the thinnings to their sold destination. A third thinning usually produces some saw logs and is usually economically viable, with the final clear fell producing the economic return. I am not an advocate of this system and far prefer the continuous cover options that eliminate the need for clear felling forests and the environmental impact that occurs as a result of this.

On sites that are susceptible to wind throw, thinning is often not encouraged. Many plantations also are not thinned as the economics of the first and second thinnings are questionable. These unthinned plantations that the forester often sees as poorly managed are a great source of poles for roundwood timber framing. The unthinned plantation allows for slow grown poles that are drawn up with minimal knots. These poles will be slow grown and growth rings will be close together. Poles of 13 metres in length can often be found where the base diameter is only 250mm. These are the softwood

LEFT MAIN European larch in early spring.

ABOVE LEFT Larch plantation greening up in spring.

ABOVE RIGHT Plantation of western red cedar and hemlock behind.

poles the roundwood timber framer is looking for; slow grown, tall and thin with closely spaced growth rings.

An increase in roundwood timber framing could remove the need for the uneconomic pattern of first and second thinning and could create a high value market for a forester who grows slow grown poles in a plantation that requires minimal maintenance. As a builder who does not use preservative impregnation in timber, species choice is very important, not just for the structural strength of the round poles but also for their durability without artificial preservatives.

PLANTING WOODLANDS FOR THE NEXT GENERATION'S TIMBER FRAMED HOUSES

The choice between planting a new woodland or allowing it to naturally regenerate is always a difficult decision. If we allow natural regeneration to occur, the woodland will evolve naturally but may not contain species which will be useful for building houses for future generations.

Traditional nurse crops, where one species is planted to draw up another and then felled, may have a useful role, especially if the plantation is left unthinned or thinned very minimally so as to create very tall thin softwood poles of a good age. Suitable combinations of plantings could be western red cedar and European larch, or Lawson cypress and European larch. All are useful species in roundwood timber framing. Sweet chestnut can be planted with European larch. Larch makes a good nurse for newly planted sweet chestnut and draws up long straight poles. Again I would watch and observe development before any thinning. The end result can evolve into sweet chestnut coppice, so it could be a sensible nurse crop within a long term vision of how the woodland would develop in the future. Douglas fir and European larch is another traditional mixture which, if left to grow on at a close stocking rate, could again produce some very good quality poles for roundwood timber framing.

One species I would recommend planting in Britain would be *Robinia pseudoacacia* (black locust). This is an amazing tree and ideal for roundwood timber framing. It is as durable as oak, fast growing with a small amount of sap wood. It also fixes nitrogen, and is a good tree for bees. I would recommend a nurse crop plantation of *Robinia pseudoacacia* and European larch. As with sweet chestnut, once cut the *Robinia* would form a coppice. It is important to keep a high stocking rate, so as to draw up the *Robinia* stems as they do not naturally grow very straight. The European larch nurse crop should help with drawing up straight poles during the first cycle.

It is time to move away from planting trees based on the market price at the current time. New woodland plantings need to be focused on the social needs of the local population as well as the larger environmental benefits of tree planting. Timber for buildings, fire wood and craft produce will become the major needs from our forest resource in the future. The mistakes of planting thousands of hectares of forest that ends up as pulp and the cost of felling, extraction and haulage that makes the forest enterprise an economic loss must surely be over.

Black locust in front of plantation of western red cedar and western hemlock.

Chapter 3

Tree Species for Roundwood Timber Framing

KEY

Guide to durability of using roundwood timber in its natural (non-chemically treated) form for roundwood timber framing

Very durable	Ideal for all external roundwood construction
Durable	Suitable for external roundwood construction
Moderate durability	Suitable for semi-protected external construction
Non-durable	Suitable for internal use in roundwood construction
Perishable	Not suitable for roundwood construction

Western red cedar

Thuja plicata

Type: Softwood.

Distribution: Native of the Pacific north west of the United States of America, but can be found north of California into Canada. Planted in the UK as a plantation timber.

Silvicultural Characteristics: In the right conditions can grow up to 75m. Will tolerate alkaline soils. Avoid frost pockets. Casts dense shade.

Timber Characteristics: The common name comes from the reddish brown heart wood; the sap wood is pale and narrow. When used untreated for roofing shingles or cladding the timber will lose its colouration and mellow to a silver grey. Wonderfully aromatic.

Strength: Western red cedar is not a strong wood for construction. It would be best avoided for construction beams, except where there are regular supports on any span. It is prone to be brittle and its bending strength is poor in comparison to the other suggested species in this selection.

Durability: Durable.

Density: Low density 390 kg/m^3.

Working Quality: Good to work, stable, coarse, straight grained, light in weight. Soft and brittle nature can cause splintering. Very fast drying.
Use in Roundwood Timber Framing: Shingles or shakes, cladding, flooring, studwork, roundwood rafters, supported structural elements.

MAIN Majestic trunk of western red cedar.

OPPOSITE ABOVE Western red cedar shingles being fixed on to the roof at Lodsworth Larder.

OPPOSITE BELOW Western red cedar floorboards, sweet chestnut roundwood frame.

Scots pine

Pinus sylvestris

Type: Softwood.

Distribution: Throughout Europe and north Asia. Native tree to Scotland. Found as far south as the Sierra Nevada in Spain and as far north as Norway.

Silvicultural Characteristics: Will grow up to 30m in height. Succeeds in most soil types including dry heathland. Thrives well on light sandy soil. Tolerates many poor soils. Establishes fast and then slows to moderate growth rate.

Timber Characteristics: Heart wood is pale yellow brown to red brown in colour. Sap wood is creamy yellow in colour and narrow in slow grown areas. Resinous.

Strength: Strong and reasonably hard, improves when grown slowly in more northern areas.

Durability: Non-durable.

Density: 510 kg/m^3.

Working Quality: Works easily and cleanly but large amounts of resin can affect quality. Takes fixings easily. Knots prone to fall out during drying. Fast drying.

Uses in Roundwood Timber Framing: Poles for internal structure, wall plates, ridge, and for stud work and flooring.

LEFT Individual Scots pine.

TOP Unthinned Scots pine plantation, producing suitable poles for roundwood timber framing.

ABOVE LEFT Scots pine plantation.

ABOVE RIGHT Individual Scots pine.

European larch

Larix decidua

Type: Softwood.

Distribution: Naturally found high in the European mountains but now planted extensively throughout Europe.

Silvicultural Characteristics: Grows up to 45m. Deciduous. Grows fast at a young age. Best quality timber is grown at altitude.

Timber Characteristics: Heart wood is red to reddish brown in colour, sap wood light and narrow. Extremely resinous.

Strength: Hard and heavy timber. Strong in construction and bending. Best softwood tested for construction strength in the round.

Durability: Moderate durability.

Density: 540 kg/m^3.

Working Qualities: Reasonably good results in machining timber. Must be pre-drilled prior to nailing and knots may drop out on drying. Can spring off the saw mill during conversion.

Uses in Roundwood Timber Framing: Most useful softwood for roundwood timber framing, due to its inherent strength and durability. Excellent strength results in bending and compression tests in the round. Can be fast grown in the south of England, so it is important to check growth rings before using for frame construction. Frame construction, cladding, floorboards, shingles, joists, stud work, rafters.

TOP Slow grown larch poles, ideal for roundwood timber framing.

MIDDLE Roundwood timber frame constructed from European Larch.

BOTTOM Larch floorboards.

TOP Larch frame in construction.

ABOVE Larch cladding.

INSET Milling up larch cladding on a mobile sawmill.

MAIN Larch tree in autumn finery.

Douglas fir

Pseudotsuga douglasii

Type: Softwood.

Distribution: Native to the United States of America but now planted extensively throughout Europe.

Silvicultural Characteristics: Grows up to 45m and when grown well produces a clean cylindrical stem. Prefers free draining sites at low elevations. Avoid chalk and limestone soils. Casts heavy shade.

Timber Characteristics: Heart wood is a light red brown colour with the sap wood being distinctly light. Some resin.

Strength: Hard. Stiff timber with good strength in bending and compression.

Durability: Moderate durability.

Density: 530 kg/m^3.

Working Qualities: Very stable. Dries well with little checking or distortion. Works well when machined or by hand. Stability makes it a useful joinery timber.

Uses in Roundwood Timber Framing: Useful timber for roundwood timber framing construction. Useful as a framing timber and for studwork and cladding, joists and rafters.

LEFT Douglas fir plantation.

BELOW LEFT Douglas fir roundwood timber frame under construction.

RIGHT Douglas fir floor joists.

MIDDLE RIGHT Stairs using Douglas fir treads.

BELOW Five raised Douglas fir cruck frames.

European ash

Fraxinus excelsior
Type: Hardwood.

Distribution: Europe, North America and Western Asia.

Silvicultural Characteristics: The tree can reach heights of up to 40m and grows best in sheltered conditions on deep calcareous soils. Grows very well on chalk and limestone and self seeds freely. Late coming into leaf and causes low shade for species growing below it. Avoid exposed sites. Avoid frost pockets and very heavy clay soils. Coppices well.

Timber Characteristics: No colour distinction between sap and heart wood. Dark hardwood is occasionally found; this is known as olive ash and can be prized by furniture makers.

Strength: Tough wood similar to oak in strength, but with high resistance to shock loading.

Durability: Non-durable.

Density: 710 kg/m^3.

Working Qualities: Ash is a good timber to work, it cleaves well and responds well to machining.

Uses in Roundwood Timber Framing: Non-durable so must be used internally. Useful for wind bracing due to high resistance to shock loading. Better strength when faster grown. Also good for flooring.

MAIN Coppiced ash at Prickly Nut Wood.

ABOVE Ash windbrace being fitted into wall plate.

INSET TOP Ash tree in leaf.

INSET BOTTOM Ash felled at Prickly Nut Wood ready to be milled.

Sweet chestnut

Castanea sativa

Type: Hardwood.

Distribution: South west Europe, Australia, North Africa and Asia.

Silvicultural Characteristics: Can reach a height of 30m or more. Often grown as a coppice crop which produces good quality poles for roundwood timber framing. Grows best on warm, sunny sand loam slopes. Avoid chalk, limestone and frosty or wet sites.

Timber Characteristics: Heart wood is yellowy brown in colour and distinct from the narrow band of sap wood. Chestnut usually only carries three years of sap wood hence making it an excellent pole for roundwood timber framing.

Strength: Good strength, resembles oak in much of its performance but not as strong.

Durability: Very durable.

Density: 560 kg/m^3.

Working Qualities: Works well, cleaves well when grain is straight. Older chestnut is often prone to spiral grain. Always pre-drill before nailing. Good durable versatile wood.

Uses in Roundwood Timber Framing: With such a small amount of sap wood, good construction strength, natural durability and usually sourced from a coppiced system, sweet chestnut finds itself at the top of my list for timbers suitable for roundwood timber framing. Useful for framing, cladding, shingles, wind braces.

TOP Sweet chestnut poles ready for framing.

ABOVE Sweet chestnut corner post with oak waney edge cladding.

MAIN: Forty year-old sweet chestnut coppice being felled for roundwood timber framing.

INSET TOP Raised sweet chestnut roundwood frame.

INSET BOTTOM Cleft sweet chestnut lath awaiting lime plaster.

European oak

Quercus robur
Quercus patrea

Type: Hardwood.

Distribution: Throughout Europe and into North Africa and Asia.

Silvicultural Characteristics: Can reach up to 30m in height. Slow growing. Often grown as a standard in coppice or within a plantation. Better timber is produced on heavier soils where there is less likelihood of shake. Avoid frost pockets.

Timber Characteristics: Wonderful broad rays are an attractive feature in the timber. Heart wood is yellowy brown and the sap wood is distinctly lighter. In younger trees the sap wood can be significant.

Strength: Very good strength properties, suitable for construction work.

Durability: Very durable.

Density: 720 kg/m^3.

Working Qualities: Moderate workable timber. Works green with sharp hand tools and machines moderately well.

Uses in Roundwood Timber Framing: Due to the large amount of sap wood when at a pole stage, oak is unsuitable for roundwood timber frame construction. However, its history in traditional timber framing and its inherent strength make it useful for underfloor ties in the frame construction. It is also the most chosen species for peg manufacture and also used in flooring, shingles and external cladding.

CLOCKWISE FROM TOP LEFT Roundwood frame showing sawn oak underfloor tie beams; oak floor and oak shelving at Lodsworth Larder; oak standards growing in coppiced woodland (volume of standards needs reducing); oak cladding at Lodsworth Larder.

ROUNDWOOD TIMBER FRAMING 31

Black locust

Robinia pseudoacacia

Type: Hardwood.

Distribution: Native to the eastern USA but now planted widely across parts of Europe and in Asia and New Zealand.

Silvicultural Characteristics: Deciduous tree which can reach a height of 80m. Tolerant of most soil types, but avoid very wet sites. Root system used for stabilising soil on hillsides. Fast growing, suckers freely, coppices. Branches have thorns. Legume family so fixes nitrogen through ability of symbiotic bacteria in the root nodules.

Timber Characteristics: Wood is of a greenish yellow colour with reddish brown veins. Narrow sap wood is distinguishable. Sawdust greeny yellow in colour and strong smelling. Larger stems are often attacked by the heart rot fungi. Growing smaller diameter round poles is a sensible option.

Strength: Very heavy and hard, rivalling the best oak, shock resistant.

Durability: Very durable.

Density: 720 kg/m^3.

Working Qualities: Moderate to difficult. Coarse texture, pre-drill before nailing.

Uses in Roundwood Timber Framing: Stem formation is often not very straight but chosen stems can be used for curved crucks. Used for frame construction, sills, pegs (traditionally used for pegging ships in the USA in the 1800s), flooring, cladding.

ABOVE Foliage of black locust.

LEFT Milling up black locust for cladding.

ABOVE LEFT Black locust growing in Lodsworth, West Sussex.

ABOVE RIGHT Black locust breaking leaf.

RIGHT Black locust in flower.

Lawson cypress

Chamaecyparis lawsoniana

Type: Softwood.

Distribution: Grows on the USA's Pacific coast from south west Oregon to north west California. Has now been widely planted in the UK, Europe and New Zealand.

Silvicultural Characteristics: The tree can grow up to 60m, has a strong conical crown and is very shade tolerant. Prefers deep soils, mild climate and plenty of rainfall. Propagated by seed or cuttings taken in spring.

Timber Characteristics: Heart wood is yellowy white to pale yellowy brown with a pale white sap wood. Ages to a silvery grey.

Strength: Very strong for a softwood. In tests in the USA, it out-performed all the other cedars and redwood. Results showed it to be 45% stronger than redwood and western red cedar in impact bending and 30% stronger in crushing strength.

Durability: Durable.

Density: 530 kg/m^3.

Working Qualities: Very good working qualities with hand or machine tools. Fine texture and straight grain.

Uses in Roundwood Timber Framing: Very suitable species for frame construction as it combines strength, durability and good working qualities. Traditionally used in Japanese architecture and boat building. Used for frame construction, decking, flooring, cladding and doors.

OPPOSITE Lawson cypress growing in unthinned plantation, ideal for roundwood timber framing.

RIGHT Very straight Lawson poles ready for framing.

BELOW LEFT Steam bending curves in Lawson cypress to use as roof rafters.

BELOW RIGHT Lawson cypress pole in the steamer.

ANOTHER POSSIBILITY

Another possible timber that may prove useful for roundwood timber framing is the swamp cypress (*Taxodium distichum*). Swamp cypress could be a useful species to plant in sheltered wetlands and areas where alder thrives. The combination of using common alder as a nurse crop for swamp cypress could be a useful planting combination for the future. Swamp cypress is hardy and frost tolerant. It has been used in many forms of construction and has good natural durability. Specimens have grown well in the UK in arboretums and I have heard of a new planting in east Devon.

Jose Ignacio Soto/Shutterstock

Chapter 4

Tools for Roundwood Timber Framing

In this chapter, I show the tools that we use — powered by humans, electricity and oil. Power tools can speed up the building process but come with many hidden costs. Every apprentice who builds with us learns to do the job with hand tools before using a power tool. This way the knowledge of hand tools and their role in our future is continued.

Cutting Tools

CHAINSAW

A chainsaw is by no means the only tool for felling a tree or cross cutting a branch, but it is quick and very efficient. Using a chainsaw safely takes practise; it should only be used with proper protective clothing and felling only undertaken after a certified course.

Once you have completed the training, the versatility of the saw becomes apparent. My winters are spent felling coppice, selecting poles for craftwork and roundwood timber framing. The chainsaw is almost used on a daily basis.

Once we begin framing the saw is used regularly for cross cutting timbers and for specific jobs within each build. A chainsaw can be adapted to become a large chop saw by use of a simple angle iron clamp known as the beam machine. This tool can be very useful when cross cutting the cruck feet on a finished frame. With a further adaption it becomes the log master, using a

toothless chain as a drive for the rotating planer blades. It is possible to peel, dig out and carve all with the same tool. This can be particularly useful for making roundwood gutters, amongst other products.

I also use an 18 volt lithium iron Makita battery chainsaw. It might look like a toy chainsaw, but it is very effective and useful in all sorts of applications, from roughing out through to more detailed carving.

PANEL SAW

A panel saw is the most readily available carpentry saw and has a wide range of uses. Panel saws are used in roundwood timber framing, in particular for cross cutting poles and during the shaping of shoulders on tenons. There are many mass produced panel saws and, if possible, a traditional panel saw with a blade that can be resharpened is a preference but if you are using the non-resharpenable version, then those with fleam teeth are the most popular to use with roundwood.

JAPANESE SAW

The Japanese make a whole range of unique tools and their saws are some of my favourite tools. The saw cuts on the pull as opposed to a conventional panel saw that cuts on the push. The blades are so fine and the cuts so clean that the wood looks like it has been planed rather than sawn. I use these saws for detail work, both cross cutting and rip sawing. They are particularly useful for rip sawing down uneven grained tenons and for flush cutting pegs.

DEBARKING SPADES

A debarker is a long handled tool with a sharpened edge which is manually used to remove the bark from a roundwood pole. The removal of bark is important as any remaining bark creates a habitat for insects and moisture, as well as making it harder to see the timber for joiner purposes. Manual debarking has been shown to produce a stronger pole than machine debarking which does not follow the natural form of the timber. The following table shows

ABOVE LEFT Makita 18 volt battery chainsaw.

ABOVE RIGHT Japanese saws.

OPPOSITE TOP Using a chainsaw to cut sweet chestnut coppice.

OPPOSITE BOTTOM LEFT Log master attachment on Husquawk 357XP.

OPPOSITE BOTTOM RIGHT A large chainsaw attatched to the beam machine becomes a chop saw.

Debarking spade peeling sweet chestnut log.

Debarking sweet chestnut logs within the coppice.

comparison in bending strengths in Scots pine. Manual debarking also allows the roundwood timber framer to choose unusual shaped poles with character and peel them, keeping their integrity intact. We also use drawknives for debarking.

Scots pine	fm p12=480 kg/m^3
Hand debarked	52.1
Partly sawn and hand debarked in tension zone	49.8
Machine rounded	48.9
Partly sawn in tension zone and hand debarked	45.0
Square sawn	43.0

The effect of processing method and form of specimen on the bending strength.
Figures from *Round Small Diameter Timber for Construction* (See Appendix D).

DRAWKNIVES

Drawknives are a wonderfully versatile tool for working green wood and have a number of specific uses in roundwood timber framing. They are used in peeling poles where a more detailed finish is required in contrast to the coarseness of the debarker. They are used for flattening off a face on a roundwood rafter, where one flat face is needed for level attachment of batons. They are used for creating flats on a roundwood beam which are mortised for use in a mortise

and tenon joint. They are also used for working a piece of oak to size before going through a rounding plane to make a peg.

ROUNDING OR ROTARY PLANES

These wonderful tools originated in use for chair making. In roundwood timber framing, they are the main tool used to make pegs of an exact diameter. The plane is rotated like a giant pencil sharpener over the selected piece of oak and the resulting peg is of an even diameter. The peg is then wedged to secure it in position. The planes come in different sizes both metric and imperial. The main peg size we use is 1 inch or 25mm; the ¾ inch and ⅝ inch are used for different joints within the frame.

LEFT Making an oak peg on a shaving horse using a drawknife.

BELOW LEFT Sorby drawknife.

LEFT Working a flat face on a roundwood rafter using a drawknife.

ABOVE Selection of rounding planes.

ROUNDWOOD TIMBER FRAMING 41

MORTICERS

When I built the Woodland House, we cut out every mortise with chisels. We didn't even use a brace and bit to remove excess wood; it was a slow methodical method of maul on chisel. On every build I do now, new trainers always cut their first mortise the same way. I believe it is important to know how to do the job with hand tools before using machinery. That being said, the chain morticer is a great tool when you have a lot of mortises to cut. Although clamping the tool onto roundwood can be more challenging than onto a 6 x 6 beam, it certainly facilitates the speed of mortising.

AXES

The broad axe is the traditional axe for hewing a timber beam out of a roundwood log. With roundwood timber framing, the need to square off a whole face is rare, and the need to square off four faces is very rare, so the skill of the broad axe is rarely in use. The main axe that is regularly in use is the side axe. Flat on one face and bevelled on the other, with a slightly cranked handle and available right or left handed, this axe is perfect for flattening off a small area and removing surplus timber without digging deep into the main beam. It is also used in roughing out a piece of oak prior to draw knifing and being turned with a rounding plane to make a peg.

ABOVE Chain morticer cutting out a mortise on a Douglas fir beam.

LEFT Right handed side axe.

CLEAVING FROE AND ADZE

Both these tools are used for riving or splitting timber. I use a froe for splitting 3-5 inch poles, for making shingles and lath. I use the small hand cleaving adze for splitting hazel for wattle and daub or panel infill.

LEFT A froe.

ABOVE A froe in action, cleaving sweet chestnut.

RIGHT Cleaving hazel with a small cleaving adze.

CHISELS

The framing chisel is constantly in use during the build and in particular during the early stages of framing. Like most tools, chisels become a personal choice. You will find a tool that sits just right in your hand, and a tool whose edge responds to the sharpening stone you use. Just like an old friend, you are straight back to where you left off (when you last saw one another). A good quality framing chisel should be an heirloom. It should last maybe three generations, unless they are all full-time framers. Therefore it is possible to find good framing chisels second hand. The main framing chisels I use in roundwood timber framing are:

Standard Framing Chisels
Long, thick blade designed for working with deep

LEFT Finishing the surface of a joint with a slick.

ROUNDWOOD TIMBER FRAMING 43

ABOVE LEFT Chiselling out a butterpat joint.

ABOVE RIGHT Using a gouge on the scribed curve of a joint.

RIGHT The chisel table keeps chisels upright and ensures at the end of the day they are all back present and correct!

mortises and able to take heavy blows from the maul.

Slick

The slick is a long handled chisel which is used like a plane to smooth off a flat surface and for finishing large tenons. I use both a 2⅜in and 3⅜in but it is the 2⅜in that gets the most use.

Gouge

A gouge is a curved chisel. I use gouges for a number of joints on the frame. Many of the joints need a curved edge as one piece of roundwood fits snugly into the curve of another.

Corner Chisel

Corner chisels are used to ensure an accurate right angle is kept when chiselling out a mortise.

DRILLS AND AUGERS

For the most extreme drilling I use a two stroke petrol Stihl powered drill. This has the power to drill through most timbers and is particularly useful when you are harnessed to the building drilling out where the crucks meet the ridge pole. The best drill we have used for high torque, slow speed drilling is the Mafell HB1E. This is my favourite all round drill for boring through roundwood. Battery technology is improving and I soon hope to use a high powered battery drill that will have the torque and control to drill some of the larger holes.

As we don't use draw pegging in roundwood timber framing, we drill through both parts of the timber in one go. This can be challenging in some cases when the two pieces of timber clamped together can be the equivalent of drilling through nearly one metre of wood. The key to successful drilling is secure clamping and good quality, sharp auger bits.

There are plenty of cheap auger bits on the market, but the Japanese ship augers and tri-bits often marketed as 'woodowls' are far superior.

The investment in good quality auger bits will save a lot of cursing during the build.

Scots Eye or Bar Augers are the hand powered alternative and will do the job perfectly well, provided they are properly sharpened. Clean holes of the exact diameter ensure good quality joints when wedged pegs are used.

ABOVE Author using Stihl 2-stroke drill.

BELOW Bar or Scots Eye augers.

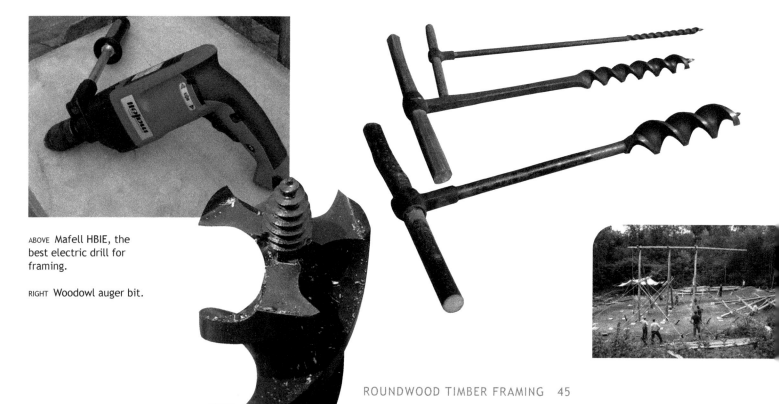

ABOVE Mafell HB1E, the best electric drill for framing.

RIGHT Woodowl auger bit.

ROUNDWOOD TIMBER FRAMING 45

Moving Tools: The Transporting of Timber

The first tool for moving the poles will be for transporting the timber from the forest to the building site. A tractor with a grab and a timber trailer can move most of the poles needed for a building, in one easy journey. As we source our timber from as close as possible to the finished building site we rarely need to engage a timber lorry for transportation.

Timber poles were always traditionally transported using a team of horses, and the use

RIGHT Extracting poles using a horse.

BELOW Mechanical extraction using tractor and timber crane.

LEFT Author moving a larch butt onto the sawmill using a mule.

BELOW LEFT Turning a sweet chestnut pole using a cant hook.

BELOW RIGHT Cant hooks being used to help stabilise feet of cruck poles during a raise.

of horses within our woodland is beginning to return. Horse adapted trailers with a grab can do the work of a tractor and can equally well deliver the poles to the building site.

MANUAL MOVING – MULES, CANT HOOKS AND TIMBER TONGS

The mule is a simple yet effective mover of timber. Constructed as a steel frame above two heavy duty wheels, it works by lever action. The simple lever action lifts the timber pole or butt off the ground by finding a fulcrum point of balance. Chains or ratchets can also be used to give added support to the timber pole. This one tool will save you many hours of struggle.

Cant hooks are a metal hook with wooden handle, used in many aspects of timber moving and framing. They are used for turning and

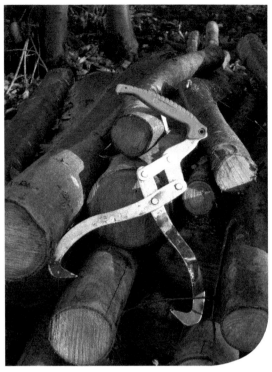

rolling timber and for releasing timber from a stack. I also use them when raising frames to give extra stability in holding the feet on the padstone (see Frame Raising). A practised user of a cant hook can do the work of three people.

Timber tongs are made of two small curved pieces of metal that are used to lift or drag smaller poles. Using timber tongs and lifting properly, bent at the knees with a straight back, will help increase the years you are able to work physically in roundwood timber framing.

WHEELS AND ROLLERS

Sets of wheels are commonly used to move timber beams around a building site. A simple welded frame with a pair of wheels will have many uses in facilitating the moving of timber and frames.

We use rollers welded to a metal frame to help facilitate the moving of large roundwood poles. I ratchet strap rollers onto the tie beams in order to assist the smooth lifting and positioning of the wall plates.

ABOVE LEFT The hook and spike of a cant hook.

ABOVE RIGHT A set of timber tongs.

LEFT Wheels for moving poles.

BELOW LEFT A roller helping to facilitate movement of a pole.

Clamping Tools

CARPENTER'S STOOLS OR TRESTLES

You simply can't have enough trestles on a building site when you are framing, they are constantly in use as portable work benches throughout the build. Every build the Roundwood Timber Framing Company undertakes involves apprentices making new trestles. This not only gives us more trestles to use on site but gets possibly apprehensive apprentices underway on a project and their hands on the tools. There are many designs for trestles but the one I have illustrated makes very good and strong all-round trestles for all applications in roundwood timber framing.

A pair of carpenters stools.

SHAVING HORSES

A shaving horse is the traditional vice of the greenwood worker. It is an all-in-one seated vice where you can sit and peel timber in relative comfort. There are many different designs and adaptations for specific uses. I use a basic model as shown here for whittling down oak to create a peg blank ready to be turned into a finished peg with a rounding plane.

To Make a Basic Shaving Horse

Cleave a straight log 20-25cm diameter and about 1.5m long. I use sweet chestnut as I have a large supply of it. It is very durable and my shaving horses live outside in a paddock! If you are moving your shaving horse around a lot, a lightweight timber maybe more suitable. Western red cedar is very lightweight and durable and can be useful for this purpose. Cleaving your log will be best achieved using wedges. If the log cleaves well you will have two shaving horse beds. Remove any excess fibres, level the cleaved log with a side axe and finish with a large draw knife. You can lighten the bed further by removing some of the excess timber on the

Views of a basic shaving horse.

ROUNDWOOD TIMBER FRAMING 49

CLOCKWISE FROM TOP LEFT Cleaving out the bed using wedges; side axing the bed; cleaving the frame with a froe; the cleaved sides of the frame; Japanese sliding bevel; hollow shoulder planes; hinging a working board.

underside of the log, but I prefer to keep the horses in the half round as it gives greater depth for the leg holes.

Create a centre line down your log using a 'Chalk Line' and mark the position for the rear legs about 5cm each side of the centre line and 15cm from the end of the log. The back legs need to be drilled so they splay outward and back. A sliding bevel could help with the angle here but I tend to do this part by eye. I do not drill right through the bed, as I like the legs a little loose, so that they can be easily removed for transportation.

Follow your centre chalk line to the other end of the bed and mark 15cm from the end of the log, the position for the front leg. This should be angled slightly towards the front of the bed.

The next stage is to make the frame. Find a piece of chestnut or ash (or other available wood) about 7.5cm diameter and about 750cm long. This then needs to be split in two. A froe is an ideal tool for this and is used in many aspects of greenwood working and roundwood timber framing. Split the piece in two down the centre using a froe and a cleaving brake. This is done by driving the froe into the round piece of timber and then using the froe by levering down on the handle whilst the timber is in the cleaving brake. If the cleave starts to run off, away from the centre, place the fatter half of the split downwards in the cleaving brake, put your hand in the split and exert pressure downwards whilst levering with the froe. This will encourage the split to run downwards back towards the centre of the timber.

Once you have two halves, clamp them together again and then mark the position of the three holes to be drilled. You may choose to make more than one centre hole, and this will allow you to set the frame at different heights upon the bed for different pieces of work.

To drill the top, middle and bottom holes I use a 25mm auger bit. This can be done with a brace and bit, bar auger or power drill and bit.

To make the top vice grip you will need a log about 8cm diameter and 40cm long. Measure the width of the bed and mark this onto the log. Use a draw knife and if you have a hollow shoulder plane, whittle down the tenons each side to fit through the 2.5cm round mortises (holes) in the frame. Repeat the process for the foot rests but use a long log about 55-60cm to ensure there is ample room for the feet on each of the tenons once it has been fitted to the frame. Next, drill a 2.5cm hole through the bed. This is usually positioned about 15cm in front of your knees when sitting on the bed, but you can drill two or three holes to allow for different positioning of the frame on the bed for different sizes of people. This drilling is best done with a level, or a very good eye, as making it horizontal is essential to ensure a well balanced shaving horse. The wooden dowel that holds the frame can be turned on a pole lath or by using a rounding plane or hollow shoulder plane. One end should be left with a head attached so that the pin cannot slide all the way through the hole. Line the frame up over the bed and tap the peg through the central hole until the frame sits comfortably attached to the bed. It should pivot freely on the central dowel. Once you are happy with the shaving horse, peg the frame. Pegs can be hand whittled and tapped into pre-drilled holes.

Next attach the working board. This sits on top of a wedged log and is traditionally pegged to the bed. I find the use of a butt hinge more flexible for a variety of uses and now prefer to fix my working board this way.

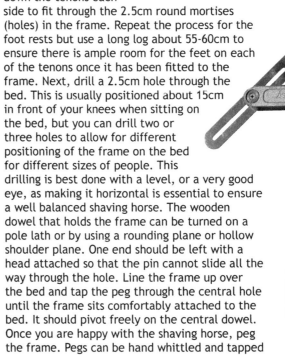

RATCHET STRAPS

I never realised how versatile the modern ratchet strap is. We use these straps for securing timbers to the framing bed or to trestles while marking and jointing the timber. These are used throughout the roundwood timber framing process and for an average build at least 30 good quality straps are necessary. The key is to find straps that are strong enough to pull the round timbers tight together before drilling, but are not so large that they are too cumbersome in relation to the size of the timbers. Don't be tempted by poor quality, at least 1.5 ton of pressure is needed. These straps are also used to secure timber to the framing bed.

CLAMPS

A wide range of clamps are used during the framing process, but the S-clamp or the deep throated clamps are the most useful.

DOGS

Timber dogs are shaped metal pins that, when hammered into two roundwood poles, hold them in place like a giant staple. They are regularly used for temporarily securing timber either to the framing bed

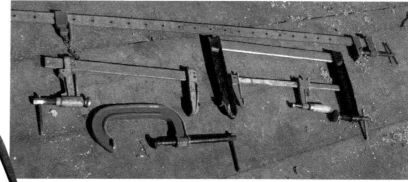

TOP Ratchet straps are essential in roundwood timber framing.

ABOVE Selection of clamps.

LEFT Timber dogs.

ROUNDWOOD TIMBER FRAMING 51

or to another log. The large dogs and smaller puppies are an essential part of the framing toolbox.

SUPERJAWS

A mention here for superjaws must be made. I have an original 'Elu' superjaws and some of its more recent successors. These are the best portable vices I have come across. They are three legged, so stand firm on uneven ground. Roundwood jaws are available for improved holding of round timber and there is no end of uses that these tools are adapted for.

FRAMING PINS

Traditional framing pins are used for marking for draw pegging. As roundwood timber framing doesn't draw peg, framing pins are used to temporarily hold together frames while final joints are being finished. They are usually about 5mm smaller in diameter than the size of the holes drilled for pegs. They can be made by anyone with basic welding skills.

ABOVE Superjaws, old and new.

MIDDLE Roundwood jaws on the superjaws.

BELOW LEFT Roundwood jaws clamping chestnut.

FAR LEFT Framing pin in use.

Measuring and Marking Tools

LEFT Mini level.
BELOW Leica Disto D8.

LEVELS

Spirit levels are used for numerous jobs on the build, especially when moving from roundwood to sawn wood where we are incorporating a contemporary stud wall into the building. Small levels are used when making flats in the same plane, prior to cutting mortises in a wall plate.

Building levels are used for foundation layout, these are particularly important with the padstone design as the padstones are usually at a range of different heights to one another.

LASER LEVELS

Modern technology has created some great tools and a quality laser level is one of them. We use three types of laser level. These are a right angle level which is useful in laying out a framing bed, foundation pads, and a crosswire laser level which is used for finding the correct angle for cutting the feet of the frames so that they sit flush on the padstones. I also use a Leica Disto. The D8 version has a lot of useful features for roundwood timber framing. The height measurement allows me a very accurate measurement of standing trees and is an advance on a clinometer. The distance measurement is incredibly accurate and with the built in screen and zoom feature it is possible to pick a point 80 metres away and get millimetre accuracy. It also measures roofing angles and does a range of Pythagorean and trigonometry calculations; all compatible to your laptop via Bluetooth if you so wish.

ABOVE Using a crosswire laser level to mark the feet for cutting.

Builders level used for padstone surveying.

LEFT Chalk line pinged on sweet chestnut pole.

INSET Chalk line and chalk refill.

BELOW Selection of squares with Japanese flexible square top.

CHALK LINE

The simple chalk line is one of the most used tools in roundwood timber framing. The ability to 'ping' a line of chalk down a wobbly roundwood pole and from that have a datum line from which you can measure and triangulate to other timbers in the frame, is essential. Wall plates are chalk lined to ensure mortises all line up.

CARPENTER'S SQUARES

2in, 4in, 6in, 12in and 18in squares are all used within the construction of a frame. These are particularly useful for marking out tenons. I find the Japanese flexible squares extremely useful for bending on a round pole.

3-4-5 TRIANGLE

A three sided triangle with sides measuring three, four and five or their multiples always produces a right angle. For example a triangle with side 30cm, 40cm and 50cm will create a right angle. These can be easily made from straight sawn timber or a pre-made folding type

can be used. Very useful in foundation layout, flooring and studwork.

LOG SCRIBERS

Log scribers are used to transfer the contoured pattern of one log onto another. They are a set of points with two spirit level bubbles attached. This allows for accurate marking, enabling a

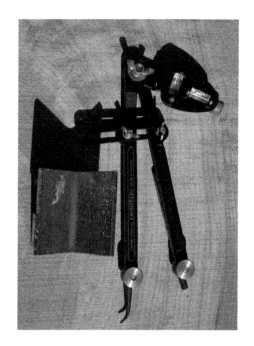 

ABOVE LEFT Log scribers.

ABOVE RIGHT Log scriber in use transferring the profile of the top log onto the one beneath.

LEFT Profile gauge being used to transfer curve of cruck.

tight and natural looking finish when the two logs come together. Used in conjunction with scribing pods, the scribing of logs now forms the basis of the main joints in roundwood timber framing. The tradition of scribing logs comes from traditional log builders and a wide range of scribing tools are available in the USA and Canada.

PROFILE GAUGES

Profile gauges are used to transfer the profile or curve from one surface to another. Used in roundwood timber framing often when flooring, the profile gauge gives an accurate scribe to ensure a good fit of floorboards to a roundwood post or cruck.

Lifting Tools

TRIPODS AND SPLAY LEGS

When I first started raising frames, especially with buildings constructed in the forest, I would lash together three roundwood poles to form a tripod, or a pair to form splay legs. Now I use load tested tripods constructed of steel for the main lifting and lighter weight aluminium tripods for lifting beams up and down during the framing process.

SNATCH BLOCKS, SHACKLES AND STROPS

I use an industrial graded snatch block (hook and pulley) for lifting frames. This, attached to a tripod, will allow a smooth action with minimum friction as the cable passes from winch to timber frame, and ensures a safe and gradual lift.

Load tested shackles are used to join winch cables to strops or to join strops together.

I mainly use 'round sling strops', these are colour coordinated as to their load bearing capacity. Damaged strops should be discarded.

WINCHES

I always use a manual winch for lifting. The slow speed of a manual winch, which works by ratcheting the handle, allows for instant stopping without any jerking of the cable or frame. It is also very quiet, and when raising frames I like total silence, so I can hear the groans of the timber and any calls of concern from the raising team.

The 'Turfor' winch is the benchmark to which all other manual winches seem to aspire. I do most raises with a 3.5 ton version but I have raised smaller frames with a 1.6 ton.

One very useful winch I use for adjusting is the 24 volt battery 'Warne Pullzall'. This is a very useful winch for shifting frames by small

TOP Steel tripod.

ABOVE Yale two-ton 360 degree swivel chain. Block and green pin load tested snatch block.

MAIN Large steel tripod with chain block attached.

ABOVE Shackles.

LEFT Selection of round sling strops.

BELOW 1.6 ton Turfor style winch.

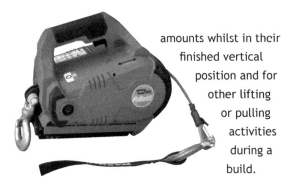

amounts whilst in their finished vertical position and for other lifting or pulling activities during a build.

CHAIN BLOCKS

Chain blocks are in constant use during the framing process and save large amounts of human lifting. We use them for lifting beams on the framing bed and for raising the jowel posts and wall plates.

ROPE

Rope plays an important role in all roundwood timber framing builds. Primarily used for guy ropes when the frames have been raised, these guy ropes form a front and back line when raising a frame. An average four cruck frame building will use nearly 200m of 16mm guy rope.

SHARPENING

In order to work successfully with wood, sharp tools are essential. I use a slow rotating water wheel 'Tormek' to sharpen my chisels and gouges. When you have 30 chisels to sharpen at a time, the Tormek can be set to the ideal angle and each chisel can be sharpened in a couple of minutes. With a water cooled grindstone, the stone carries the water continuously to the grinding surface. This cools the surface and eliminates the risk of overheating that can occur with dry bench grinders. After grinding, a burr will develop on the upper side of the edge. This burr is then honed off with either a fine slip stone, or with the Tormek — I use a leather honing wheel.

The other sharpening stones I use are Japanese water stones. These soft stones come in a variety of 'grit' sizes and can produce a very fine edge.

TOP LEFT Warne 24 volt battery winch.

TOP RIGHT Chain blocks suspended from tie beams for hauling up jowl posts.

BELOW LEFT A raised frame is secured with guy ropes.

BELOW RIGHT Tormek water cooled grindstone.

Chapter 5

Construction

Foundation pads of york stone ready to receive the frame.

INSET TOP Digging one cubic metre foundation pits.

INSET MIDDLE Compacting aggregate infil with 'jumping jack' whacker plate.

INSET BOTTOM Padstone layed on recycled crushed concrete infil.

Foundations

On most building sites, foundations involve major earth works, levelling of the soil, extra drainage and can cost up to one third of the total build cost. With roundwood timber framing, we use the natural contours of the land, creating foundation pits without the need to level the site. This approach minimises soil disturbance and is a very economic approach to foundations. Roundwood timber frame buildings can be built directly onto slopes as well as level ground.

Ten years ago, I spoke to people about the need and opportunity to construct foundations which are not made of fresh concrete. Say the word 'foundations', and people are already conjuring up the image of a concrete footing. The only concrete I have used in our buildings has been at the Pestalozzi International Village build, where we used recycled crushed concrete to form the base for the padstones. Roundwood timber framing does not use fresh concrete. Instead we create foundation pits on average of 1 cubic metre and fill these with crushed local aggregate and then place a padstone on top.

LAYING OUT

We use the traditional laying out system of batter boards. These are constructed from short boards nailed to stakes which are driven into the ground. The batter boards are set to a right angle using a 3-4-5 triangle. These right angles

ABOVE Line drawing of roundwood timber frame showing foundation pits and padstones, dotted line represents ground level.

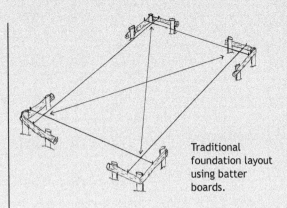

Traditional foundation layout using batter boards.

make the extremes of the foundation and help to keep the building parallel and in square. With the main footprint marked out, it is possible to measure and mark the positions of the all the padstones ready for excavating the pits. A small roundwood timber frame like the one at Pestalozzi International Village had only 12 foundation pads; others like 'The Diddlers' had 52. Each foundation pit is dug carefully with a mini digger. On average we dig a cubit metre of soil out of the hole. The edges are squared off with a spade. We then back-fill with a local aggregate, compacting the aggregate at regular intervals before adding more with a jumping jack compactor. Once the pits are fully compacted a York stone slab is laid on top and bedded in with sand. Levels are checked to ensure the padstone is level. Once the padstones are level then it is time to take level measurements off each padstone. One padstone is chosen as the datum stone and the measurement established from the datum stone is used to calculate all the measurements for the building.

USING A BUILDER'S LEVEL

Two people are needed to operate a builder's level. The level needs to be recalibrated annually to check accuracy. It is set up in a position where it is possible to oversee the foundation site on a tripod. The level has adjustments and its own levelling bubble to ensure it is set up in a level position before you start taking measurements.

Once the builder's level is set up in its position the tripod should never be moved and care must be taken not to jog the builders level on its tripod stand. It will now swivel from side to side without going out of level so you can survey the whole building plot.

Next you need to choose your datum padstone. I tend to choose either the highest or

OPPOSITE TOP Using a builders level and staff to survey padstone levels.

OPPOSITE BOTTOM Measurements and calculations to establish differences in padstone levels.

MAIN Ensuring a padstone is level and then using string lines with the aid of a 3-4-5 triangle to set the building square on the foundation stones. Where the strings cross will mark the centre of the post.

lowest padstone in the landscape as it makes the maths easier. All other padstones will then be either higher or lower in measurement from the datum stone, rather than taking a mid height padstone where some measurements will be lower than the datum and some higher. This allows a greater risk of human error than choosing the lowest or highest padstone as your datum. The second person stands with the measuring staff. They position the measuring staff on the padstone whilst the first person operating the level finds the measurement in millimetres by looking through the viewer on the level at the staff. The viewer has cross hairs that can be adjusted in brightness to ensure an accurate measurement. This datum

ROUNDWOOD TIMBER FRAMING 63

measurement is taken and then the process is repeated on all other padstones. This gives a series of figures in millimetres. The differences in figures are then calculated from each padstone measurement compared to the datum. For example, if padstone number seven is 1041mm and the datum is 1011mm there is a difference of 30mm in level between the datum point and this padstone. The 30mm must go into the calculations on the framing bed for measuring the extra length in cruck needed below the floor level.

DAMP COURSE CONSIDERATIONS

As the process of roundwood timber framing has evolved, the contact point between wood and stone at the bottom of the feet has always been an area about which I have been concerned. As each building has been constructed, we have improved techniques to create a damp proof layer between the two. On the Woodland House, the feet sit directly on the padstones. Although the posts are chestnut and therefore very durable, over time the end grain will absorb moisture through contact with the padstone. My plan for the long term with the Woodland House is to jack up any suspect feet, cut a few inches off the bottom and replace them with a sacrificial wooden block. This could then be replaced at an interval of 40 years or so, whenever inspection shows this to be necessary. Granaries and barns have for years been built on staddle stones or stone piers to support timber posts and there is little difference in the roundwood timber frame approach. However, if it seems likely that the base of the feet may need replacement, it makes sense to minimise the regularity of when that becomes necessary. More recent builds have involved painting the base of the feet with a lacquer of natural ashfelt and linseed oil. This is a natural product that gives a rubberised finish and seals the end grain of the timber. Added to that we now make sure that there is 5mm of slate trimmed to the shape of the post to ensure that any water arriving on the padstone does not work its way to the base of the post. This 5mm lift above the padstone, plus

ABOVE LEFT Slate inserted under timber feet prior to being trimmed to shape.

ABOVE RIGHT The base of a cruck post which has been painted with natural ashfelt and linseed oil to seal the end grain and stop ingress of moisture.

ABOVE Traditional staddle stones hold up a granary.

FAR LEFT Timbers rest on shaped stone piers.

LEFT Brick and timber is used to lift main post above moisture level, The Weald and Downland Museum.

the natural lacquer and a good air flow beneath the building that this design allows, should ensure a good lifespan before any feet need their bases replaced.

EARTH ANCHORS

On windy and exposed sites we have secured the buildings to the ground using 'Earth Anchors'. These are steel anchors with a cable attached, which are hammered into the ground using a compressor and pneumatic hammer. They're hammered about 3-4 metres into the ground, then the cables are tensioned up. The force tensioned into the cables is measured in kilonewtons. As the cable is tensioned, the anchor opens in the soil and pulls against the weight of the soil on top of it. The end of the cable is then attached to the frame to anchor it to the ground.

LEFT The gauge measures the force of the tensioned cable with attached 'Earth Anchor' in kilonewtons.

RIGHT An MR2 model earth anchor attached to a stainless steel cable prior to hammering into the ground.

Frame Construction

Roundwood timber framing can be used to construct most conventional forms of timber frame but the most common design I have used is the cruck frame, or more appropriately in roundwood timber framing, the A frame. This design lends itself to roundwood timber framing; the frame aesthetic of the round pole is appreciated as it elegantly extends in graceful simplicity throughout the building. This design is simple but very structurally sound as the A frame creates immense triangulation throughout the building. It also allows a lot of flexibility into which partition walls, if so desired, can be placed.

Box frames are equally possible using roundwood construction, however although

LEFT Using a compressor and pneumatic hammer to knock the anchor up to 4 metres into the ground.

66 ROUNDWOOD TIMBER FRAMING

ABOVE Roundwood box frame, prior to inserting windbraces.

LEFT Roundwood box frames with temporary bracing, used for raising the frames.

they gain more internal usable space, much of the roundwood aesthetic is lost and they are therefore rarer in roundwood buildings constructed so far.

THE FRAMING BED

The framing bed is a key part of roundwood construction and the first timber creation that takes place on any new build. The bed is made from either floor joists or roof rafters, 6in x 2in. Pairs of 6in x 2in are nailed together (putting washers on nails will make removing them after the frame is finished an easier prospect). These pairs are supported on wooden posts that are dug into the ground. The pairs of 6 x 2s are fixed to the posts so that they are level with one another and then extra pairs are secured, as a second layer running at right angles to the first layer in a grid pattern. Spend time ensuring your framing bed is level and in square, as it will save you much time and you will not have to deal with problems further into the build. This second layer of 6 x 2s is laid out carefully to represent key positions of the frame you are creating. So with an A frame, one of the pairs will be positioned to mark where the tie beam will be and another pair will mark the floor height. Each frame can then be laid on the bed to mirror the previous frame. The details that you have calculated from the levels on the padstones will give the length of each log below the floor level, so that each frame can have the legs cut whilst resting on the framing bed at the correct height, and they all rise up level. A round log of similar diameter to your chosen ridge pole can be positioned as a template so that each cruck is jointed and the ridge pole sits comfortably in the horns of the jointed frame.

The framing bed is such a key element in roundwood timber framing. I particularly like the use of joists and rafters in its construction.

TOP: Laying out a roundwood frame on the framing bed.

ABOVE Layed out cruck frame ready for jointing.

LEFT: Checking the framing bed is 'in square'.

OPPOSITE Using the floor joists as a framing bed to make the verandah frame at Lodsworth Larder.

Once the frames have been completed and raised, the framing bed is dismantled and the framing bed timbers then find their final resting place as joists or rafters in the very building that was jointed on top of them. Once the building is up and the floor joists are in place, the timbers can form another framing bed for extra frames, such as for verandahs and decks. Most of the key decisions and experimentation has taken place whilst timbers are laid out on the framing bed. It is a place of construction, evolution and learning.

Sway braces secured from cruck to cruck stop racking.

LEFT: A sway brace, showing detail of tenon and sculptured shoulder.

BELOW LEFT Wind braces from jowl post to wall plate alleviate racking.

BELOW RIGHT Curved sway braces help resist racking between frames.

BRACING

Whatever type of roundwood timber frame you choose to construct, bracing will play a significant part. Bracing ensures the building stays straight and does not, over time, begin to lean in a particular direction. Most of the bracing that goes into roundwood timber frames, we refer to as wind braces. These are braces that span at 45 degrees usually between jowl posts and wall plates. When exposed they form an attractive feature, as well as carrying out

ROUNDWOOD TIMBER FRAMING 71

the essential job of stopping the building from racking. Other braces we use I refer to as sway braces, as these form cross bracing from one cruck to another. They are usually exposed and can form a stunning feature in an open plan building where their unique shape becomes an interesting component of the visual frame.

UNDERFLOOR SUPPORTS

I have devised two different underfloor support systems in roundwood timber framing so far. In Lodsworth Larder, underfloor tie beams join the crucks and allow attachment of the jowl at any chosen position along the underfloor tie beam. The underfloor tie beams are then joined with joists running between them. The other system involves an underfloor support beam that spans the length of the building and joins the cruck frames together. This beam is jointed into the cruck frame with a bird's mouth, and then the underfloor support beam is sandwiched between the cruck post and the jowl. The jowl then continues up to meet the tie beam above. In some examples, where there is a considerable fall in the landscape and therefore some variation in construction angle between the lowest and highest A frame, a cranked jowl, commonly known amongst roundwood timber framers as a 'disco jowl', is introduced. This ensures the underfloor support is sandwiched along its length but cranks around it to pick up the wall plate above, so that it runs parallel to the other wall plate, and roof layout is therefore not over-complicated. Where underfloor support beams are used running the length of the building, it will be necessary to introduce at least one or more (depending on span width of building) extra underfloor support beam(s) on staddle posts, to support floor joists running across the whole width of the building. The staddle posts are attached to the underfloor support with mortise and tenon joints.

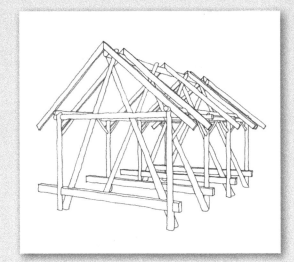

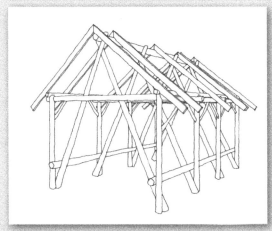

FROM TOP
Frame design with underfloor tie-beam.

Frame design with underfloor support beams.

Bird's mouth cut in cruck, ready to receive underfloor support beam.

LEFT Underfloor support beams supported at 3 metre intervals with staddle posts.

BELOW Underfloor support beam secured between cruck and jowl post.

BELOW LEFT Underfloor support system showing additional mid-span support.

ROUNDWOOD TIMBER FRAMING 73

WALL PLATES

In many roundwood timber frames the wall plates are located on top of the jowl posts and the tie beam. This creates a very secure wall plate where the three all come together. With a roof pitch of 45 degrees, the available second floor space will be limited as the roof rafters will inhibit access to the outside of the building on the second floor. By jointing the jowl post to the tie beam and extending it another metre above the tie beam, then securing the wall plate in this position, the second floor space becomes larger, allowing access across the whole width of the building. As with all roundwood timber framing it is important to remember that all poles taper. Wall plates around 13m long may have quite a taper along the whole length and it might be necessary to wang the wall plate across to meet all the tenons on the jowl post. This can be done with ratchet straps, a lightweight winch or tourniquet.

OPPOSITE Wall plate resting on tie beams prior to being lifted onto awaiting tenons on extended jowl posts.

OPPOSITE INSET Jowl tenon where wall plate meets jowl post and tie beam at the same level.

ABOVE Top wall plate fitted, lower wall plate awaiting positioning on tenons.

LEFT Wall plate suspended above jowl post with chain blocks.

ROUNDWOOD TIMBER FRAMING 75

Roundwood Timber Framing Joints

To cut any of the following joints successfully, please practise on spare timbers before beginning on your chosen framing poles.

LEFT Chiselling a tenon on a roundwood staddle post.

BELOW a flat formed on a sweet chestnut pole and a mortise chiselled out.

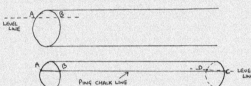

Chalk lining a pole prior to forming flats for mortises.

MORTISE AND TENON JOINT

This traditional timber framing joint is used in a number of locations in the roundwood timber frame. It is used on the top of wooden staddle posts supporting underfloor support beams and on top of jowl posts where they engage with the wall plate. These tenons are on average 4in high x 4in long x 2in wide. A flat is formed on the wall plate and then the mortise is cut out. To form the flat on the wall plate, place the wall plate log on the framing bed or on top of two pre-levelled trestles. Draw lines A to B and C to D with a spirit level at the same height, measured off the framing bed. This height will be determined by the diameter of the wall plate. The aim is to draw the lines, so that the minimum amount of timber need be removed from the wall plate. The golden rule is: 'Never remove more than one third of timber'. Chalk lines are then pinged between points B and C and points A and D. Timber is removed from between the chalk lines where a mortise must be cut for the jowl post tenon to be inserted. The mortise can be marked out using the mid

76 ROUNDWOOD TIMBER FRAMING

RIGHT The blue chalk lines run the total length of the log, allowing for the option of creating a flat for a mortise at any point along its length. The pencil level line marked on the end grain ensures a reference point if the log moves during jointing.

BELOW A finished tenon. Note chamfering on the leading edges to ensure no ripping out of grain on contact with the mortise.

point between the two chalk lines as the centre line of the mortise. The tenon must have a flat shoulder and the measurements for the tenon are made using a small right angle square, taking measurements from the end of the pole which will have been cut at a right angle to a chalk line pinged along its length. The wall plate is then lowered onto the tenon on top of the jowl post. Chain blocks are used for this process. It is important to ensure the shoulder of the tenon joint is flat and meets the flat on the wall plate snugly. Also, make sure the tenon is slightly shorter than the depth of the mortise; this will ensure that when the timbers shrink, the end of the tenon is not being forced by the bottom of the mortise. These mortise and tenon joints are pegged with a ⅝in peg.

WIND BRACES

Wind braces also use a mortise and tenon joint. We use a wooden box known as the 'Magic Meseg Box' named after Rudi, one of the framers from the Roundwood Timber Framing Company who constructed the first and subsequent boxes. These boxes are a jig, designed to give a constant 45 degree angle on the shoulders of the wind brace to ensure an exact fit to the wall plate and jowl. The box has an opposite template to offer to the wall plate and jowl. This ensures the angles are accurate. The length of the box and the template can be increased or decreased depending on the length of wind brace needed, but the angle stays constant.

The box is an adapted mitre box, and ensures the angles of the shoulders and plain of each of the tenons match. Something that is easy to get wrong with roundwood.

The chosen piece of wood for the wind brace is pinged with a chalk line (to have a central line to work from) and is placed in the box; chocks are used to ensure the pole sits in the right position. The angled ends of the box are the mitre angles for the shoulders, and the tenon can be measured out and accurately cut. The template is used to mark out the exact positions for the mortises to be cut. Once the mortises have been cut, the wind brace can be offered into position. Once any adjustments have been made and all wind braces are in place, the wall plate can be lowered for the last time and two ⅝in pegs inserted through each mortise and tenon joint.

The 'Magic Meseg Box' with roundwood windbrace inserted ready for mitre cuts.

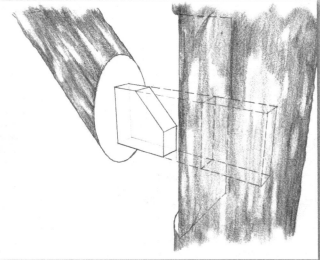

LEFT Rudi using windbrace template to mark mortise positions.

ABOVE Line drawing of windbrace mortice and tenon.

Wall plate being lowered onto jowl post and windbrace tenons. We have fitted up to eight windbraces on any one wall plate, with five or six jowl tenons to line up with mortises — many hands are helpful.

ROUNDWOOD TIMBER FRAMING 79

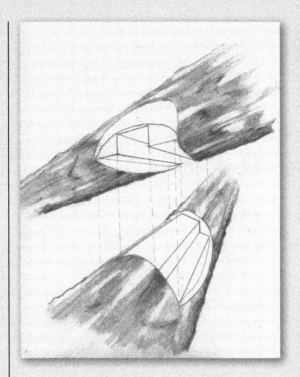

FAR LEFT Line drawing of cogged cruck joint.

LEFT Chiselling out the joint.

BELOW Both parts of the finished joint.

COGGED CRUCK JOINT

This joint is used where the two crucks meet and cross over one another to form the horns into which the ridge pole rests. Once the orientation of which cruck overlaps the other has been established, the under cruck is fixed to the framing bed with ratchet straps or timber dogs and remains fixed in this position throughout the process of scribing and cutting the joint. The upper cruck is lowered using tripods, to sit on a scribing pod, directly above the other cruck. A scribing tool is then checked on a levelling board and used to transfer the profile of one log to another. With the scribe lines in place, mark level lines on the under cruck by measuring up off the framing bed. These lines will form the heights of the shoulders and the top of the cog. I usually recommend making the cog about one inch higher than the shoulder and 1½-2in wide depending on the diameter of the cruck pole. The cog is then chiselled out. The upper cruck is turned over and secured to be worked on, and the flat is formed, working from between the scribe lines.

Once level, the upper cruck is turned back over and lowered onto the cog of the lower cruck. If they are not totally flush, adjust the flat face on the upper cruck. Once they fit flush, mark the cog profile on the flat face of the upper cruck. Turn over the upper cruck and chisel out the mortise for the cog. Using a fish tail gouge, chisel out the profiles of the scribe lines until the two crucks sit together flush. This joint will be secured with a one inch seasoned oak peg, which will be wedged both sides.

OPPOSITE CLOCKWISE FROM TOP Using a spirit level to transfer level lines from the framing bed; scribed log showing transfer lines in pencil; lowering one log onto the other; line drawing of butterpat joint; snug fit of the butterpat joint; view of the butterpat, note slight dovetail.

80 ROUNDWOOD TIMBER FRAMING

BUTTERPAT JOINT (OR DOVETAILED NOTCH)

This joint is used to join both the cruck to the tie beam and the jowl post to the tie beam. This joint was named on a build after it resembled a pat of butter sitting in a butter dish. More technically, its origins lie in the square notch joint as used by many log builders, but due to the angle of the cruck where it meets the tie beam and the extra security of the dovetail added to lock it in place, this joint has been designed and adapted for this specific purpose in roundwood timber framing. The butterpat is always cut into the tie beam. The tie beam is lowered using tripods and strops to sit over the crucks. A scribing pod can be used and the profile of the cruck is scribed onto the tie beam. Using a spirit level to come vertically off the framing bed, level lines are marked on the tie beam to form the top of the butterpat. The butterpat is chiselled out and dovetailed to further help lock the joint. The tie beam is lowered onto the cruck. The mortise part of the joint is marked off the butterpat and then cut into the cruck using the scribe lines to follow the profile of both logs. The joint is secured with a 1in seasoned oak peg which is wedged both sides.

ROUNDWOOD TIMBER FRAMING 81

HYBRID COGGED DOVETAIL JOINT

This joint gained its hybrid name because it is the only joint used where round poles meet squared timber. In the building of Lodsworth Larder, we used sawn underfloor tie beams; this was so that we could hang joists between them on joist hangers rather than cut into a round underfloor tie beam which would raise the height of the floor further. By using a sawn underfloor tie beam in this instance, it also gave a flat top and bottom surface to attach floorboards above and panel vent boarding below in order to house the insulation between them.

The underfloor tie beam is lowered down onto the cruck that is jointed onto the framing bed. Once in position the crucks are marked where they meet the sawn edges of the underfloor tie beam. These marks will form the edges of the cruck joint that overlaps the sawn underfloor tie beam. The cruck part of the joint is worked on with a full flat, formed about 1½in deep. Measurements and levels are taken off the level framing bed as with the other joints. The cogged dovetail is then marked onto the flat and cut a further 1in in depth. The sawn underfloor tie beam is then lowered onto the jointed cruck, and the mirror image of the cogged dovetail is marked onto it. The joint is then cut and held together with a 1in seasoned oak peg which is wedged at both ends.

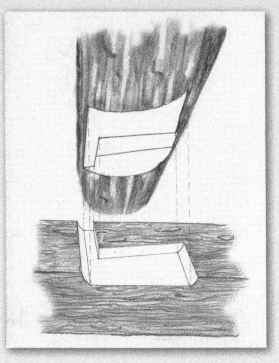

FAR LEFT Line drawing of two dimensional hybrid cogged dovetail joint.

LEFT Line drawing of three dimensional hybrid cogged dovetail joint.

BELOW Hybrid cogged dovetail joint cut into chestnut round and sawn oak beam.

If the height of floor above ground level is not an issue, then we use a round underfloor tie beam as in the 'Woodland Classroom' and use dovetailed butterpat joints.

SCRIBING LOGS

A log scriber is used to scribe the profile of one log onto another. The first step is to create a levelling board. This is a wooden board attached to a post or building, which needs to be exactly vertical. Onto this board, you draw a plum line. Once you have opened your scribing tool to the required distance for the logs you are scribing, place both tips of the scribe tool against the plum line and adjust until both levels are reading zero. Place the scribing tool so that the point is touching one log and the pencil the other and the levels are both reading zero. Keeping the levels at zero allows the scribing tool to glide across the profile of the one log enabling it to mark the profile onto the other log, keeping the levels at zero at all times. Scribing takes both practise and patience. The use of indelible pencils which mark on wet timber can be helpful in getting clear markings on the logs.

CLOCKWISE FROM TOP
Pencil scribe of log profile, note use of scribing pod; scribing the profile of one log onto another; the joint lies hidden within the scribed logs; both faces of a scribed butterpat joint; pencil marks the scribe.

ROUNDWOOD TIMBER FRAMING 83

PEGS

As pegs are used to secure the joints in roundwood timber framing, they deserve a special mention. Oak pegs are the choice for roundwood timber framing because of their historical pedigree. I used seasoned oak boards, which I then rip down into a square section slightly larger than the diameter of the finished peg. These are then drawknifed on the shaving horse before being turned on a rounding plane to create a cylindrical peg. In traditional square timber framing, draw pegs are used. These are tapered pegs that are forced through two slightly off-set holes and, as they are knocked in, they draw both parts of the joint together. In roundwood timber framing, the joints have the unusual property of often being hidden beneath the curves of the timber, making accurate draw pegging more awkward.

For this reason and for additional strength, I have always chosen wedging pegs as the best option. The joint is ratcheted tightly together and then a hole is drilled the exact diameter of the peg. The peg is hammered through the hole and then cut off, protruding about 1in each side of the joint. A Japanese saw is used to rip the slit through each of the pegs. The direction of the cut is chosen away from the ingrain and the wedges are hammered home. As the joint made from greenwood shrinks onto the seasoned oak peg, it tightens fast and the addition of the wedges ensures the peg cannot slide either way.

TOP LEFT Using a rounding plane to make a peg.

MIDDLE The ridge pole pegged to the cruck.

BOTTOM LEFT Inserting a wedge into the peg.

LEFT Author wedging a peg.

A frame can use over two hundred pegs, so many friendships are formed over peg making.

ROUNDWOOD TIMBER FRAMING 85

Frame Raising

Frame raising has always been a celebrated moment in all of the roundwood timber frames I have constructed. It brings together all the work that has been steadily progressing on the framing bed and in one day the skeleton of the building emerges. It is a long day and one throughout which I have to remain very focussed. We raise our frames by hand with the help of a manual Turfor winch. Attention to detail, particularly in relationship to health and safety issues, are at the forefront here. There is often an over exuberance of people wanting to help, a keenness to be part of the excitement of the day. Having led eight raises I have learnt that a safe raise is not dissimilar (I imagine) to a military operation. I need to know who I have for the raise, their skills and concentration levels, and then each person is briefed and allocated to a particular task. We do a rehearsal to ensure everyone knows their positions and it gives me an opportunity to sort out any problems. Once planned, no one else is allowed within the raising area, and we raise in total silence. This gives me the opportunity to listen to the strains and creaks of the frame but also to hear anyone call with a concern.

At the end of the day the revelry can begin, and many a pint will flow as the structure stands tall and proud. I have memories of every raise but some in particular stand out. Lodsworth Larder was a wonderful raise day, with about 200 villagers looking on as we raised the village shop, blessed by the vicar and adorned with children's garlands, all within the car park of the village pub. 'The Diddlers' was also a wonderful day. Chris and Lucy took the winch and hauled up the final frame of their house together, amidst an onlooking cheering crowd of friends and locals. The bonfire burnt and the celebrations went on into the early hours.

This type of emotional and significant landmark occasion could never be felt with bricks and mortar. It is the majestic trees, felled from the forest, jointed with chisels and raised with ropes that connect us to the magic of what building a house really means. They touch a latent instinct, a connection to those carpenters of the past and, as we raise another, it feels like we are setting the pattern for the future.

TOP The first frame sits with its feet resting on the padstones; it will be standing vertically on these after the frame is raised.

ABOVE The final frame is positioned, prior to the raise.

A successful day's frame raising. A five cruck frame raise is a long day. This five cruck, sweet chestnut frame we raised at 'The Diddlers' was the essence of the joy of building. Watching Chris and Lucy winch up their house was poetry in building.

ROUNDWOOD TIMBER FRAMING 87

THE RAISE DAY

Preparing for a raise takes time as there are many elements that need to be in place before the actual raising can begin. All of the frames are laid out with their feet on the padstones that they will eventually stand upon. Metal bars are hammered into the ground to stop the padstones from sliding during the early part of the raise. (These are removed and can be reused for further builds after the raise.) To start with, the frames are lifted up manually with the help of a tripod and chain block and then chocked in place with timber off-cuts to assist with the raise. It is natural that a frame lying in a horizontal position will want to drag rather than raise. To make it raise, the feet need to be anchored in position and any height gained by pre-lifting the top of the frame makes the process easier and smoother. To anchor the feet we use both a cant hook with human energy, in the early part of the raise, and rope tourniquets. The rope tourniquet has proved to be one of the most useful raising tools and, if the feet begin to slide, it can be tightened. When the frames are getting close to vertical it can be released. Each frame has four rope lines attached to it: two front and two back lines that act as guy ropes to secure the frame when it reaches it vertical position. These ropes

FAR LEFT Raised frames showing guy ropes and woodland greenery.

LEFT Metal stakes are hammered in to stop the padstones from sliding prior to the raise.

BELOW Ropes and ratchets are secured to the frames prior to raising.

BELOW LEFT Rope tourniquets help control the slide of the frame feet during the early part of the raise.

stay on the building until the wall plates and bracing are all in place.

The winch cable runs through a snatch block suspended from a tripod. I have one large load-tested tripod with adjustable legs, as so far we have not had a raise on even ground. The hook at the end of the winch cable attaches to a shackle which grips the end of a soft sling, this in turn is wrapped around the cruck. Before the raise begins I give a safety briefing and everyone takes their position. Eight people including myself is ideal. One person operates the winch. Two people on cant hooks on the feet of the crucks release the cant hooks once the angle of raise is above 60 degrees, and take up the front lines. Two people are on tourniquets. Two people are

Support posts are put in place to help raise the frames above horizontal before the winch begins to lift them. Note the ridge pole running through a roller positioned on the frame tie beam of one of the frames.

on the back lines. One person (raise leader) conducts the proceedings. This is an essential role. It ensures the lift goes safely and overall instructions and decisions must come from the raise leader, there is no time for disputes or differing opinions when you have 2.5 tons of timber delicately balanced in mid air.

A four cruck frame is easily raised in a day, a five or six cruck is possible on a summer's day, if you start early and finish late.

Once the first frame is winched up, it becomes the anchor point for the second frame. A snatch block is then suspended from the first frame. The second frame is often the most awkward. The first frame is only supported by guy ropes and yet it is used as the high point for securing the snatch block. This can mean the first frame sways a little as the second one begins its ascent. The second frame also has to pick up the ridge pole, one end of which went up with the first frame. The other end of the ridge pole, which is on the ground, needs to be manually walked round so it sits in the horns of the second frame. There is a lot of friction as the second

LEFT A tractor is positioned as an anchor point. Note the large yellow tripod in place with snatch block and cable ready for the first lift.

MAIN Raising a Douglas fir frame in the woods.

frame goes up and the ridge pole has to slide between the horns of the cruck. An occasional nudge of the end of the ridge pole, with a long ladder, frees up the movement and the second frame becomes upright. Temporary bracing is then fixed diagonally between frames one and two at this point and the most difficult part of the raise is over. Further frames should come up fairly easily and with frames one and two braced together, the fixing point for the snatch block is more stable. Provided the measurements were correct on the framing bed, the next frame should pick up the ridge pole and so become vertical.

ABOVE Some raising days are very social occasions like the raising of the village shop in Lodsworth.

When all the frames are raised and the ropes are attached, the shackles, winch and tripod have been dismantled and the final temporary braces are secured in place, this is the moment to have a beer and admire a day well spent in roundwood timber framing.

If you are not convinced, you can hire a crane!

INSET Five cruck Douglas fir frame standing proud.

BOTTOM Once the frame is raised, the beer appears and the party can begin.

Chapter 6

Beyond the Frame

ONCE THE FRAME IS RAISED, additional parts are added to it. Depending on the frame design, you may have already jointed the jowl posts and be ready to start work on the wall plates. To get the wall plates up onto the building can take considerable effort. We strap rollers onto the tie beam and using human power and a small winch called a 'pullzall', we are able to lift the wall plates to fit on top of all the tie beams. This is the perfect place to work on them and offer them up to the tenons on the jowl posts. Often the tie beam will not line up with all the tenons and needs to be wanged using ratchets and the pullzall. If the jowl posts need to be lifted into place, a chain block suspended from each tie beam and a round sling strop will make lifting them into place relatively simple.

At this point, we have constructed the scaffold. We use lifting beams which can be positioned to allow a chain block to be hung from them. This enables the wall plate to be moved up and down as it is offered to the tenons with minimum human efforts. Once the wall plate is sitting comfortably on all the tenons from the upright jowl posts, then it is time to fit the wind bracing. Using the 'Meseg magic box' we scribe the positions of the mortises using the template, after creating flats on both the jowl and the wall

LEFT Using a chain block to lift a jowl post into position.

ABOVE Chain blocks suspended for lifting inner and outer jowl posts into position, straw bale walls run between the two rows of jowls.

ABOVE LEFT A butterpat joint being used for fitting jowl post to the tie beam.

ABOVE RIGHT Butterpat joint hidden within the curves of the timber.

BELOW Wall plates in position with wind bracing in place. Note lifting beams with chain blocks for controlling rise and fall of the wall plates during jointing.

plate, so the template fits flush. The mortises are then marked and chiselled out. The wind brace comes out of the box and should fit pretty snugly into both mortises. Minor adjustment to ensure good contact between the shoulders may be necessary. Once all wind braces are in place the wall plate can be lowered for the last time. This usually involves a few people as the braces may need holding as the wall plate descends. Once in place, two ⅝in pegs are inserted through each tenon on the wind braces and 1in on the jowl post tenons. The main frame is now complete and it is time to move onto the roof.

A chain block and round strop secured to a lifting beam, makes the control and lifting of the wall plates an easy operation.

Cutting a Roof

Cutting a roof varies considerably depending on the style of building it is being installed upon, and the complexity of the roof itself. With roundwood timber framing we are now moving into more complicated roofing patterns. A gable roof, though simple in form, needs some thought as the wall plates and ridge all taper and the rafters are unlikely to be cut using a template, as they will vary down the length of the building.

One process used at 'The Diddlers' was to treat the roof in a more traditional fashion by installing a ridge board above the ridge and fixing the rafters to the board, with small amounts of letting into the ridge as the roof progressed.

My preference is to fix the rafters to both the ridge and wall plates and this involves a bit of patience in measuring and marking before roofing begins. I place a laser level on the top of the fatter end of the ridge and pick up the laser beam on a target at the far end of the ridge pole, thus calculating the difference in fall along the ridge. The ridge pole and wall plates are always the most carefully chosen poles and even over a 13m run, 50mm is usually the maximum taper. Letting a rafter 50mm into the ridge to balance up the taper and leave the roof looking good is a simple enough process, as by now you have built the frame. I measure the chosen length of the roof along the ridge and temporarily fix one pair of rafters to it and then cut in the other pair by 50mm. I check that the

'The Diddlers' — with the wall plates completed. A ridge board has been used for creating a level to work from along the ridge.

ROUNDWOOD TIMBER FRAMING 97

First rafters in place when using a ridge board for leveling.

top points of each are level and then move down to the wall plate. Again the drop in level due to the taper of the wall plate is usually less than that of the ridge, partly because we can see it and allow for the wall plate to sit slightly higher on the jowl at the thinner tapered end. With the butt end of the rafters sitting on the wall plate, I measure the distance between them and adjust them so it corresponds to the distance they are fixed apart on the ridge. I then measure the diagonals to see if they are the same and that the roof is in square (a similar process to how the foundations were laid out to be in square). If they do not match, I adjust the bottom ends and

INSET FROM LEFT Inserting 6 x 2in roof rafter into bird's mouth cut into wall plate; chiselling out bird's mouth joints in wall plate; bird's mouth cut into wall plate. Note different coloured chalk lines used to work bird's mouth positions along the whole wall plate.

recheck the measurements and diagonals until they correspond and the roof is in square. The next stage is to allow for the taper on the wall plate. This can be calculated with a laser level, as with the ridge. The bottom of the rafter needs to be bird's mouthed into the wall plate. With a sawn frame there is a fixed wall plate with a right angle on which to fix the bird's mouth. With a tapered roundwood wall plate, you have to create the virtual edges of the sawn wall plate using chalk lines. We use two different coloured chalks to show the top and bottom lines, which will be chiselled out to form the right angle upon which the rafter will drop. These chalk lines have taken in the variation of taper. Once the end rafters are in place string lines are stretched across the top of them to check the intermediate rafters are correct. I use a 'Leica Disto D8' to check the angle of fall on each rafter to make sure any discrepancy is within 0.5 degree. The rafter gaps are then measured and set up, usually at 60cm centres, for insulation purposes. Remember at this point to allow for any roof lights, which will need rafters doubling up, and the position of any flues or roof vents that may be needed within the building. The process is slower than having a template and all rafters exactly the same, but the result of cutting an accurate square roof on a roundwood timber frame gives a feeling of great satisfaction.

TOP LEFT Using string lines to check levels along the whole roof.

TOP RIGHT Rafters picking up two wall plates in cat slide roof at Lodsworth Larder.

ABOVE Rafters cut into ridge pole rather than using a ridge board.

ROUNDWOOD TIMBER FRAMING 99

ABOVE FROM LEFT Slate roof; clay tile roof; thatch roof.

ROOFING MATERIALS

Being a man who works with wood, I like to use shingles or shakes in many of the buildings. These do have an advantage of being very light, weighing six (western red cedar) to 10 (oak) kg/m². Compare this to clay tiles 35-40 kg/m² or thatch 34 kg/m². Having a lighter roof means a lighter structure in the frame of the roof. When you look at an engineer's calculation, there will be live loads and dead loads calculated for a roof. A dead load consists of the structure, batons, membranes, shingles or tiles and a live load allows for the weight of snow landing on a roof. This winter after we had

BELOW After the rafters are completed, a breathable waterproof membrane is fitted prior to batoning.

Shingling Lodsworth Larder with local Sussex western red cedar shingles. Baton and counter baton visible on top of 'permaforte' membrane.

12in (30cm) of snow on one of my very flat roofs, and the forecast was talking of a further 4in (10cm), I got up on the roof and shovelled it off. Shovelling a roof full of snow 12in (30cm) deep, made me very conscious of the live load weight of snow.

Different roofing materials can cope with different angles of pitch on a roof. If you have a very flat roof with perhaps only 6 degrees or more of pitch then you are limited to a sheeting material. This could be corrugated iron or 'onduline' (organic plant fibre and bitumen) the choice I have used on a number of roofs. If you are using thatch or shingles, then the steeper the roof, the quicker the water will shed and the longer the thatch or shingles will survive. So really a minimum of 45 degrees is needed. You can go lower on a pitch with shingles down to 30 degrees but do not expect your roof to last so long. Slates are a comfortable choice at a roof pitch of 30 degrees.

Section of twin membrane design used in Lodsworth Larder roof.

12in (30cm) of snow on the Woodland House roof.

MAIN Shingled roof complete, note fixings in place for solar photovoltaic panels.

INSET Cleft sweet chestnut shingles.

OPPOSITE LEFT Creating a shingle ridge cap.

OPPOSITE TOP RIGHT Looking up at the Lodsworth Larder verandah roof with its round wood rafters.

To ensure good insulation properties in these buildings, we are now using a two membrane system in the roofs and walls. These remove the chance of any cold air finding a way through the system. The counter baton allows air circulation under the shingles speeding up their drying out process after heavy rains. The membranes are breathable yet waterproof allowing trapped moisture to escape from the building but keeping external water out.

If your roundwood timber frame is a non-insulation building, a shed, workshop or outdoor covered space, then roundwood rafters can be used and the batons and shingles will be exposed from underneath as at Pestalozzi. At the Sustainability Centre, we steam bent all the roundwood rafters to get the desired roof curve for the 'Woodland Classroom'. A cleft shingle made from oak or chestnut will outlast a sawn shingle. The sawn shingle has more exposed end grain. The western red cedar shingles we use are not from old growth forests of the United States of America and Canada, as are most western red cedar shingles found at builders and roofing merchants in the USA, they are locally sourced and locally made from Sussex grown trees.

ABOVE Sussex western red cedar shingles stacked and air drying.

Walls

Straw bales are a perfect match for a roundwood timber frame. With a raised floor, good overhangs from the roof, and sometimes a verandah, the straw bales are well protected even before they have been lime plastered and clad. Source your straw bales the season before you are going to use them. Check they are tightly bound and let them store in a barn so that they have a chance to settle and compress over the year. You may have to pay extra for this service but it will be a worthwhile investment. Where using bales, build a flat ladder on your floor out of 4 x 2 and infill the gaps with sheep's wool or similar natural fibre insulation. Fix your first bale spikes into this ladder or, for real strength, drill through and pick up the floor joists below. The bottom row of bales is then built on the ladder. If you lay out the ladder accurately, the outside edge can be used to pick up the skirting board if so desired. This ladder allows for any flooding

TOP Straw bale walls combine beautifully with roundwood timber frames. Note one bale height to form window seat below large window.

FAR LEFT Lime plastering the bales.

LEFT A completed window seat.

or spillages and is particularly sensible in a bathroom. The bales are fixed with two spikes which go through to the lower layer; these can be made from coppiced chestnut or hazel. To give the bales a good finish, I cut the edges using either a chainsaw or an angle grinder with an arbo-tech carving blade. This gives the bales a fine textured finish. The bales then need a skim coat of lime plaster on both sides. This is applied using a hurling trowel (a trowel shaped like a small shovel that releases the plaster with a flick of the wrist). The internal bales are then given a hand trowelled three to one sand and lime putty mix with horse and cow hair and then finished with a three to one sand and lime putty plaster. Lime dries slowly and in hot weather it is best to drape Hessian in front of the plaster, spraying it regularly with water to slow down the speed the plaster dries out, therefore avoiding cracking. In cold weather, it can be the opposite. The lime plaster can seem to take weeks to go off and I have often brought in a wood burner to speed up the process. Remember to scratch each coat with a scratch comb (or a piece of wood with a few nails in it) as it is going off, and wet it down before applying the next layer. These walls can then be finished with lime wash. Each coat of lime wash is structural and adds strength to the wall as well as colour. You can make a plain lime wash with lime putty and water and then add pigments to create colour washes. The key reason for using lime on straw bales is that

Straw bale walls in place with internal walls being marked out. Note orange mineral cable for ultimate fire safety.

LEFT Clay plaster being applied to cleft chestnut lath wall.

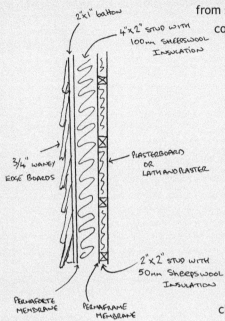

BELOW Twin membrane design used in Lodsworth Larder walls.

Lime plastering onto cleft chestnut lath.

it breathes, so any trapped moisture is able to escape through the wall, rather than becoming trapped within the bale. It is also a rodenticide.

An alternative on the inside of the bales and for internal partition walls is to use an earth plaster. This is a clay based plaster, mixed from soil and with a reasonable clay content, sand, and some form of binding. Finely chopped straw works well as a binding and it can be worked into the lath or split hazel wattle using your hands. It is a pleasure to use after lime as you don't need the protective clothing and eye goggles.

On Lodsworth Larder, we lime plastered the gable end to create a fire break for building regulations between the larder and the village pub. This was applied onto split chestnut lath. The laths were split out of usually a 4in round of chestnut and only out of the straightest part of the pole. These are split through and through (with the centre lath being discarded). It is similar to milling a log, except you are cleaving rather than sawing. The lath has a naturally rough finish and is far better for the plaster to adhere to. The gable end was plastered with a haired 3.5 to 1 mix of sand and lime putty and then two coats of unhaired also with a 3.5 to 1 mix. It was finished with four coats of natural parchment lime wash and makes a lovely contrast to the timber work all around it.

As Lodsworth Larder is a shop, space was a premium, so we decided not to use straw bales but go for a highly insulated stud wall. This consisted from the outside of ¾in waney edged oak boards. Always make sure there is a good overlap of heart wood over heart wood and remember these boards will shrink a little as

External wall at Lodsworth Larder. Note contrasting materials at roundwood corner post. Oak waney edge boarding and split chestnut lath awaiting lime plaster.

INSET Finished lime plastered gable end at Lodsworth Larder.

ABOVE Sheeps wool insulation in external wall.

ABOVE RIGHT Finished exterior of waney edged oak.

RIGHT Hemp and recycled cotton insulation fitted between floor joists.

they dry. Where the waney edge boards meet the roundwood poles, we fix a thin strip of roofing felt which is squashed between the two. We then feltch the roofing felt with boiling water and press it with a blunt bolster. The boards are fixed onto a 2 x 1 baton, then a Klober Permoforte membrane onto a 4 x 2in stud work at 60cm centres; this is filled with sheep's wool insulation. Then we add another permaframe membrane, 2 x 2 stud work with sheep's wool insulation and a plasterboard finish. Again we have used the double membrane approach to avoid any cold air bridging. These membranes should be taped at the joints. Electrical services can be run within the 2in studwork.

There are a number of other alternative natural insulations that can be used instead of sheep's wool. Natural fibres such as flax, hemp and cotton as well as wood fibre are all alternatives and the practicalities of using them can be found in my book, *The Woodland House*.

Floors

The floor system on roundwood timber framing involves either a 6 x 2in or 8 x 2in floor joist which, depending on the choice of frame, will either be let into the large roundwood underfloor supports or hung with joist hangers off the sawn oak underfloor tie beams. Either system will then need a base to hold the insulation. For this I use Panelvent. A particle board made without glues or toxic properties, it is bonded by natural resins in the wood. 9mm Panelvent is sufficient and can either be screwed onto the underside of the floor joists, if you have sufficient space under the building, or when you fix the floor joists you can add an extra 4 x 1in to the underside of the joist. This gives an inch each side of the joists and enables the Panelvent to be cut into strips and dropped in from above. The joists are laid out at 400mm centres and then the void is filled with insulation. I use Douglas fir or larch for the joists.

For the floor boarding, I use tongue and groove kiln dried timber. My preferred choice is

Tongue and grooved oak floorboards scribed around crucks and jowl posts give the impression of trees growing out of the floor.

Joists secured by being cut into underfloor support beams. Note double beams in centre to support wide span.

oak as it is so hard wearing and also ages with character over time. One of the difficulties can be finding a firm to kiln dry, plane, thickness, and tongue and groove your oak. Often this process means that even if you supply your own oak, the cost per metre of the flooring will be more than imported Euro oak boards which are currently flooding the market. Of course the cost of the imported boards does not allow for the energy cost or the timber miles. Nor does it give you the satisfaction of standing on a floor made from trees you have known. At Lodsworth Larder, the oak came from Prickly Nut Wood and was transported to the other end of the parish to W. L. West and Sons saw mills. They milled the wood through and through in January and air dried it until June. Then they kiln dried the boards and converted them into the tongue and groove floor boards. These then travelled half a mile back to the building site. In all the timber had travelled about two miles from the place it was felled to its final destination.

ABOVE Oak from Prickly Nut Wood air drying at W.L.West and sons sawmill in Lodsworth, this became the floor at Lodsworth Larder.

LEFT 8 x 2in joists being hung with joist hangers off 8 x 6in oak underfloor tie beams. Note: 4 x 1in board attached to underside of joists to support 'panelvent' and hold insulation.

110 ROUNDWOOD TIMBER FRAMING

Chapter 7

Roundwood Timber Frame Builds

The Woodland House

Description: Author's woodland dwelling. This was the prototype roundwood timber frame and is standing up well to weather and heavy snow loads.

Detail: Sweet chestnut frame, larch joists and rafters. Oak flooring. Straw bale walls. Lime and earth plasters. Warm cell insulation in roof and floor. Waney edge oak cladding. Cleft sweet chestnut shingles. Solar photovoltaic electrics. Wind turbines. Solar water and wood fuel heating. Sandstone aggregate with York stone pad foundations. Rain water harvesting. Earth anchors.

Length of Build: Seven months, built with the help of many volunteers.

Architectural drawings of the Woodland House.

114 ROUNDWOOD TIMBER FRAMING

OPPOSITE, L TO R, TOP TO BOTTOM:
3D model of the Woodland House
Foundation pits
Frames ready for raising
Working on the frame
Shingling using rope and harness
Shingling with sweet chestnut shingles
Installing verandah frame
Sunlight hits the floorboards
Warmcell (recycled newspaper) insulation being placed in the hopper
Using small diameter coppiced sweet chestnut in the roof
Contemplating the floor

ABOVE Author working on his house with model in view

OPPOSITE, L TO R, TOP TO BOTTOM:
First layers of straw bale walls
Keeping the round poles exposed by cladding between them
Diamond window in north gable
Open plan living area with rumford fireplace
Earth plaster internal wall
Freshly completed Woodland House
'Onduline' roofing on leanto bedrooms
Oak deck
Woodland house framed by chestnut trees
View leading to the bathroom

THIS PAGE Southern gable seen from the garden with the house beginning to age into its surroundings

The completed barn with timber drying area below catslide roof

The Woodland Barn

Description: Roundwood timber frame workshop.
First softwood roundwood timber frame.

Detail: European larch frame. Western red cedar wall plates. Roundwood sweet chestnut rafters. Larch floor joists. Butted oak floor. Onduline roof. Solar Photovoltaic electrics. Sandstone aggregate with York stone pad foundations. Earth anchors. Waney edge oak cladding.

Length of Build: Three months.

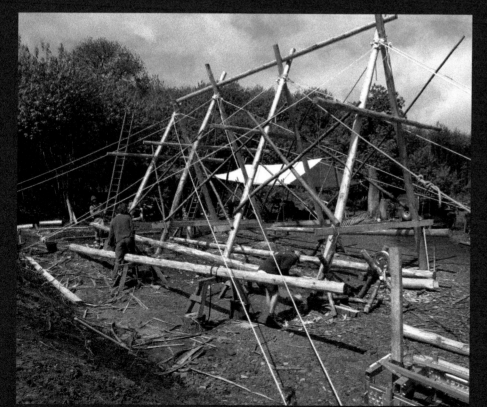
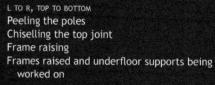

L TO R, TOP TO BOTTOM
Peeling the poles
Chiselling the top joint
Frame raising
Frames raised and underfloor supports being worked on

ROUNDWOOD TIMBER FRAMING

L TO R, TOP TO BOTTOM
Jowl posts being fitted
Sweet chestnut roundwood rafters form the catslide roof
Roundwood rafters being fitted
Fixing 'onduline' roofing sheets
Completed barn with its first snow fall

Floor joists in place and sway braces fitted in roof

Verandah frames fitted and rafters being laid out

The Diddlers

Description: Charcoal burners dwelling for Chris and Lucy Wall Palmer and their children Ember, Woody and Hazel.

Detail: Sweet chestnut frame. Straw bale walls. Western red cedar shingles. Douglas fir joists and rafters. Western red cedar flooring. Sandstone aggregate with York stone pad foundations.

Length of Build: Frame took eight weeks. Chris is still currently working on the build when he can take time off from his charcoal burning and shingle making activities.

ABOVE LEFT Many hands make light work
ABOVE First frame laid out on the framing bed
LEFT Foundations laid out and first poles peeled

ROUNDWOOD TIMBER FRAMING 123

L TO R, TOP TO BOTTOM
Frame raising underway
Wall plate and bracing
End of frame raising
Frame showing slope in the land
View of frame showing fall in the landscape
Western red cedar shingles made on site
View of finished frame with first rafters going on

MAIN
Roof completed and studwork underway

Pestalozzi International Village

Description: Horticultural barn for new garden and local food operation.

Detail: European larch and sweet chestnut frame. Roundwood chestnut rafters. Western red cedar shingles. Larch flooring. Coppiced hazel infill on hand rails. Recycled crushed concrete aggregate with council paving slab foundations. Earth anchors. Waney edge larch cladding.

Length of Build: Seven weeks.

MAIN Larch frames with sweet chestnut rafters in place

LEFT Laying out foundations
MIDDLE Peeling roof rafters
RIGHT Preparing to raise

L TO R, TOP TO BOTTOM
Framing bed constructed
Laying out the first frame
Two frames and the ridge pole raised
Raising completed
Jowl post tenon preparing to insert into wall plate mortise
Batons fixed on the roof
Weaving cleft hazel in the gables
Roof ready for shingling
Author fixing shingles

OPPOSITE
Looking up through the roof

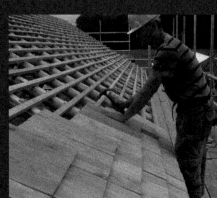

128 ROUNDWOOD TIMBER FRAMING

TOP TO BOTTOM, L TO R
Weaving hazel in the roof
The roof complete
External view of woven gable showing sway braces
Sway braces in the roof, one side shingled
Laying the floor
Woven hazel handrails
Shingles closing in
Waney edge larch boards being fitted
Internal view of completed roof with hazel

MAIN
Author in front of completed building

Hampshire Forestry Barn

Description: Forestry building in woodland.

Detail: Douglas fir frame. Douglas fir floor joists. Douglas fir roof rafters. Western red cedar shingles. Limestone aggregate with York stone pad foundations. Sawn oak cladding.

Length of Build: Eight weeks.

MAIN Moving the snatch block during raising

LEFT Timber tripod
MIDDLE Ready to raise
RIGHT First frame on the way up

L TO R, TOP TO BOTTOM
Preparing to raise
Final adjustments before raising
Five frames raised
Rafters in place ready to batton
Fitting the barn doors
Shingling the roof
Douglas ridge pole with protective oak fascia
Finished barn

MAIN
Roundwood frame visible beneath conventional sawn oak feather edge boards

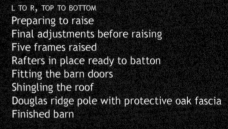

136 | ROUNDWOOD TIMBER FRAMING

Lodsworth Larder

Description: Community owned village shop.

Detail: Sweet chestnut roundwood frame. Oak underfloor tie beams. European larch wall plate and ridge. Douglas fir joists and rafters. Sweet chestnut roundwood rafters on verandah. Sheep's wool insulation. Oak waney edge cladding. Ash internal wind braces. Chestnut external wind braces. Hazel handrail panel infill. Chestnut laths. Lime plastered gable end. Western red cedar shingles. Solar photovoltaic. Sandstone aggregate with York stone pad foundations. Heat exchange system for reusing heat given off by fridges to heat the building.

Length of Build: 18 weeks.

Lodsworth Larder was awarded Best Community Building 2010 by Sussex Heritage Trust

MAIN
Contemplating where round meets sawn

ABOVE LEFT Frame laid out on the framing bed
ABOVE Fitting the wind braces
LEFT Lowering the wall plate

ROUNDWOOD TIMBER FRAMING 137

L TO R, TOP TO BOTTOM

Where three poles come together, jowl, wall plate and tie-beam
Sawn rafters on main roof
Roundwood rafters over verandah
Upper floor joists going in
Looking up at the verandah roof
The main frame prior to rafters
Floor joists fitted
Shingling the roof
Finished roof

MAIN Wall plate, tie beam and jowl all meet. Note, this windbrace has been partially flattened to ensure cladding ends up tight to jowl post

140　ROUNDWOOD TIMBER FRAMING

TOP TO BOTTOM, L TO R
Waney edge oak boards meet sweet chestnut jowl post
Cladding the northern gable
Shop fitting
Lime plaster on southern gable end
The bridge between mezzanine store rooms being constructed

Open for business
Vent for heat exchange system cut into oak waney edge cladding
2.8 KW solar photovoltaic array
The bridge lit up

ABOVE
The completed village shop

Sustainability Centre

Description: Woodland Classroom, multi-functional education facility.

Detail: Lawson cypress roundwood frame. Steam bent Lawson cypress roof rafters and wall plates. Western red cedar shingles and floor/deck boards, sweet chestnut roundwood batons, hazel handrail panel infill. Limestone aggregate with York stone pad foundations. Earth anchors, cordwood walls, adobe floor and Rumford fireplace.

Length of Build: 20 weeks.

TOP TO BOTTOM, LEFT TO RIGHT
Steam bent roof rafters setting in a jig
Author laying out steam bent roof rafters
Batoned roof with cordwood wall below
Steam bent wall plate
Steam bent sloping wall plates form the rounded ends of the building
Shingling underway
Lawson cypress frame laid out ready for raising
The curve of the steam bent rafters
Rounded shape of the roof becoming apparent

OPPOSITE
Shingles curving around the rounded roof

Extroduction

BUILDING REGULATIONS

The principle of building regulations is to ensure that buildings are constructed to a set of standards across the industry. The building regulations are regularly updated, often improving the energy efficient requirements of a new building. They also serve to check that the building will be of sound construction, meeting tested figures used by the industry. With roundwood timber framing (at present), the first obstacle to get by is the likelihood that the building inspector probably will not have come across roundwood in a building, or encountered a building where the foundations are not made from concrete. It will probably be necessary therefore to employ an engineer to run calculations on the timber you are using to justify to the inspector that the building will be structurally sound and the foundations won't sink. This is a situation that I hope will change as more structural tests are carried out on roundwood and there can be a visual grading system can be formed that will meet the requirements to pass the structure of the frames without the need to pay an engineer each time one of the buildings is constructed. Like all individuals in life, building inspectors from different areas vary considerably in their outlook on these buildings. I have been fortunate to find some very enthusiastic inspectors who are very helpful and it can be possible to have constructive conversations about the best ways to meet the building regulations schedules with a roundwood timber frame construction.

Beyond the frame and foundations, the rest of the regulations are fairly easily met. Natural insulation is very efficient and the wall systems I have described easily meet the U-values required for the building.

FIRE REGULATIONS

Fire regulations form part of the building regulations and the roundwood timber frames I have constructed have not been questioned. I am, however, often asked about the fire resistance of roundwood. Being a woodsman, I start with the property of a log to be burnt on a fire. A split or sawn log burns far more readily than a round log, hence this knowledge can be applied to roundwood poles in comparison to sawn poles. In Canada, where a lot of log building takes place, there are very detailed figures on the fire resistance of roundwood.

For a practical reference, I have to turn to the unfortunate example of Stephen Owen, whose woodworking studio at Cranleigh School in Surrey burnt down due to a tarpaulin being set alight by a work lamp, just before the studio was completed. What was interesting in the post mortem of the fire was that while the metal roof and other parts of the building burnt and melted, the roundwood sweet chestnut frames charred but remained standing.

Stephen has since rebuilt the studio and created a very interesting structure.

ARCHITECTS

Architects can be a great asset when building roundwood timber framing. Their ability to draw, their knowledge of the building regulations and ideas on utilising the space within a roundwood timber frame can be invaluable. I am fortunate to have worked with local architects who now have a good understanding of this vernacular and can design and adapt the roundwood timber framing system to create interesting designs.

I feel we are at a changing time in architecture. The days of designing a building of architectural fantasy and then sourcing the materials needed from all around the globe are coming to an end. It is time for architects

A trainee fixing floor joists at Lodsworth Larder.

to start with the available resources and subsequently design from what is available. As a woodsman, if an architect pays me a visit, needing materials to build a local house, I can tell him/her what trees of which species are available at that particular time. They can then design the house from the resources available to them locally. This type of woodsman/architect communication will lead to locally sourced sustainable buildings.

WASTE – AN UNUSED MATERIAL

Having visited many building sites, I am always astonished by the amount of waste generated on a daily basis and the huge piles of unused materials heading for landfill. There are some good recycling schemes beginning to make use of these surpluses before they reach landfill, like local wood recycling schemes. However, it is clear that the current building industry in the United Kingdom is a very high energy, high waste industry. At the end of a roundwood timber framing build, the majority of what could be considered waste is wood, and therefore can be used in the wood burner to help heat the building you have just constructed. Natural products, locally sourced, reduce timber miles and reduce landfill. We need to start changing our building industry, creating buildings that minimise energy inputs in transport and manufacture and produce no waste that cannot be recycled within the locality of the building project.

THE FUTURE

At present there are only a handful of people (as far as I know) working with roundwood timber framing. Many traditional framers and log builders are beginning to see the potential as an alternative or extra string to their bow when it comes to framing. On each build we undertake at the Roundwood Timber Framing Company, we train new framers. We take on volunteers to learn the basics of a construction process and run training courses to educate on the detail of the joints. Many of these trainees are now out building roundwood timber frames on their own, adapting the methods and passing on the knowledge to others.

I hope this book will inspire many to pick up a chisel and have a go; others will no doubt spend time on one of the roundwood timber frame courses myself and others are offering. There is a real opportunity to train a wide range of people in an aesthetically pleasing and low impact form of framing. It should be possible for people to go out into the forest, select timber and build houses for themselves and their communities. This could prove to be an essential key skill as we move into the future and need to find simple and practical solutions for the changing times we face.

There needs to be more cohesion between log builders, square timber framers and roundwood timber framers. There are huge overlaps

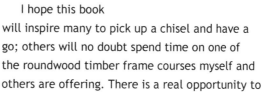

'Kasthamahal' – Nepalese for temple of wood, named by a Nepali student at Pestalozzi International Village.

ROUNDWOOD TIMBER FRAMING 147

between these disciplines and opportunities to share and develop skills.

We need more testing of roundwood, to make the process of building regulations easier and the need for engineers' reports less necessary. The T.R.A.D.A. (Timber Research and Development Agency), along with the Forestry Commission and tied in with students of engineering, could work together to find a format for visual grading and strength testing on the timbers I have recommended in this book. Architects and engineers need to become involved in the understanding of roundwood timber framing at an early stage. It is pleasing to get information from architectural students who have chosen their careers after being inspired by the Woodland House. Colleges are bringing in more modules on sustainable building and roundwood timber framing is becoming a component in some of these.

We need to think carefully about the trees we are planting now and their uses when they mature in the future. As our climate warms, our landscape will change and our children will not thank us for the trees we plant now if they do not mature into the materials they will need to build their houses in the future.

Above all, the intention in sharing these ideas with you is to inspire you to get involved and learn to work with your hands. The future we are facing needs practical hands-on skills and these key skills, such as managing our forests and building our dwellings from the timber, are the skills of a sustainable future. I am inspired by the enthusiasm of many young people I meet, who are saying no to a traditional career path and instead are learning life skills and returning to the land. May there be many more who take this path and pass on their knowledge to those being born today.

Author (blue hat) giving instruction during raising and preparations at Lodsworth Larder.

APPENDIX A

LODSWORTH LARDER
ENGINEERING REPORT ON ROUNDWOOD TIMBER
FRAME BUILDING (extracts from Adams Consulting
Engineer's Report)

IT IS PROPOSED TO CONSTRUCT A TIMBER CRUX FRAME HOUSING TWO FLOORS TO PROVIDE A SHOP FOR THE LOCAL COMMUNITY WITHIN LODSWORTH.

THE MAIN FRAMES WILL COMPRISE PREDOMINANTLY 250mm DIA SWEET CHESTNUT WITH A 150x200mm DEEP TIE MEMBER SUPPORTING FLOOR JOISTS SPANNING BETWEEN THE FRAMES.

AT GROUND FLOOR LEVEL THE FLOOR JOISTS ARE CARRIED ON JOIST HANGERS WHEREAS AT FIRST FLOOR THE JOISTS ARE NOTCHED INTO THE TOP OF THE CROSS MEMBERS.

FLOOR JOISTS ARE 50x200mm. DEEP HOME GROWN DOUGLAS FIR AND ARE SPACED AT NOMINAL 400mm. CENTRES. ON THE GROUND FLOOR A 100x25mm. DEEP PLATE IS PROVIDED ON THE UNDERSIDE OF THE JOISTS AND THE PROJECTION SUPPORTS THERMAFLEECE INSULATION PANELS LAID BETWEEN THE JOISTS.

IN ADDITION TO THE SELF WEIGHT THE GROUND FLOOR HAS BEEN DESIGNED TO SUPPORT A LIVE LOAD OF 4kn/M² AND THE FIRST FLOOR A LIVE LOAD OF 3kn/M² TO ALLOW FOR NOMINAL 1M OF STORAGE SPACE.

THE FIRST FLOOR JOISTS ARE FIRE PROTECTED BY TWO LAYERS OF PLASTERBOARD. HOWEVER THE TIMBER FRAME IS PREDOMINANTLY EXPOSED AND THE PRINCIPAL MEMBERS HAVE BEEN CHECKED FOR A 1Hr. FIRE PERIOD AND A CORRESPONDING CHARRING RATE OF 0.5mm/Min.

LONGITUDINAL STABILITY IS PROVIDED BY DIAGONAL BRACING BETWEEN THE FRAMES ABOVE FIRST FLOOR AND BY A SERIES OF KNEE BRACED FRAMES WHICH SUPPORT THE RAFTERS ALONG THE PERIMETER OF THE BUILDING.

ALL JOINTS ARE TRADITIONAL CUT MORTICE AND TENON OR HALF LAPPED AND PEGGED USING HARD WOOD DOWELS.

GROUND FLOOR

	kg/m²	Service Load DL	LL
9mm T&G OAK BOARDING	13 ⎫		
50mm×200mm JOIST @ 400°/c	12 ⎬ 30	0.3	
100×25mm BOARDS	2 ⎪		
40mm THERMAFLEECE (25kg/m³)	3 ⎭		
DESIGN LIVE LOAD (SHOP)	—		4.0
Σ DESIGN LOADS – GROUND FLOOR		0.3	4.0

FIRST FLOOR

	kg/m²		
9mm T&G OAK BOARDING	13 ⎫		
50×200mm JOISTS @ 400°/c	12 ⎬ 53	0.52	
100mm THERMAFLEECE (25kg/m³)	3 ⎪		
25mm PLASTERBOARD KKIM	25 ⎭		
DESIGN LIVE LOAD	—	—	3.0
Σ DESIGN LOADS – FIRST FLOOR		0.52	2.5

NOTE:
DENSITY OAK = 690 kg/m³
DENSITY DOUGLAS FIR = 490 kg/m³

GEN STORAGE – OWING TO REDUCED HEADROOM EITHER SIDE OF THE CENTRAL WALKWAY STORAGE LIMITED TO APPROX 1m.

ROOF

	kg/m²	Service Load DL	LL
RED CEDAR SHINGLES	5 ⎫		
50×25 BATTENS @ 600°/c	1 ⎪		
50×25 COUNTER BATTENS	1 ⎪		
100mm THERMAFLEECE	3 ⎬ 38	0.37	
50×50 BATTENS @ 600°/c	2 ⎪		
50×150 RAFTERS @ 600°/c	6 ⎪		
PHOTOVOLTAIC ARRAY (SAY)	20 ⎭		
SNOW LOADING	—	—	0.6
Σ DESIGN ROOF LOADS		0.37	0.6

MATERIALS

FRAMES – SWEET CHESTNUT   OAK (D30)
$P_{bc} = 7.6\ N/mm^2$ $9.0\ N/mm^2$
$E_{mean} = 11,300\ N/mm^2$ $9,300\ N/mm^2$
Density = 510 kg/m³   690 kg/m³

FLOOR JOISTS – DOUGLAS FIR (HOME GROWN)
$P_{bc} = 6.2\ N/mm^2$ ⎫
$E_{mean} = 11,000\ N/mm^2$ ⎬ SS GRADE
Density = 490 kg/m³ ⎭

WIND LOADING FOR BUILDING @ WALSWORTH
LANDRANGER POSITION: SU 926 229
ALTITUDE TAKEN AS ~ 100m

BUILDING – TYPE FACTOR, K_b		BS 6399-2
$K_b = 1$		TABLE 1
BUILDING HEIGHT ≈ 6.25m		
$C_r = 0.02$		FIG. 3
∴ BUILDING WITHIN LIMITS OF APPLICABILITY		
$V_b = 21.3\ m/s$		FIG. 6
$V_e = V_b \times S_b$		2.2.3.1
$S_b = 1.6$		TAB. 4
$V_s = V_b \times S_a \times S_d \times S_s \times S_p$		2.2.2.1
⇒ $V_s = 21.3 \times 1.1 \times 1.0 \times 1.0 \times 1.0$		
⇒ $V_s = 23.4\ m/s$		
⇒ $V_e = 23.4 \times 1.6$		
⇒ $V_e = 37.4\ m/s$		
$q = 0.613\ V_e^2$		2.1.2.1
⇒ $q = 0.86\ kN/m^2$		

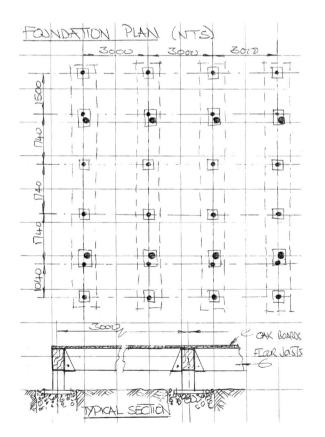

FOUNDATION PLAN (NTS)

TYPICAL SECTION

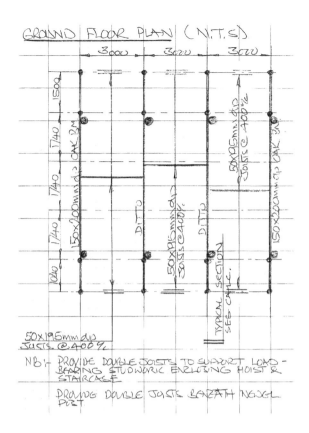

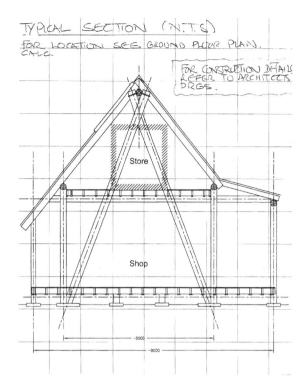

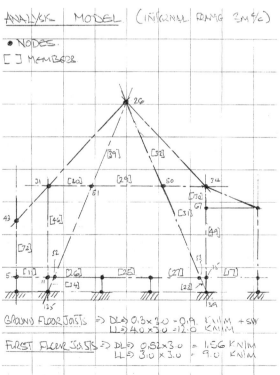

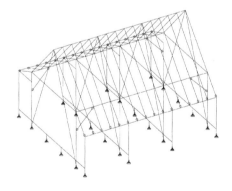

Whole Structure

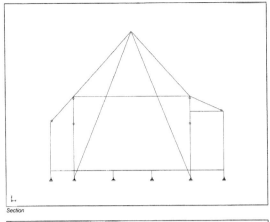
Section

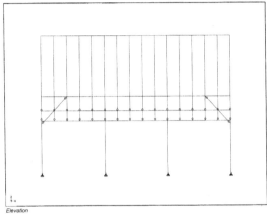
Elevation

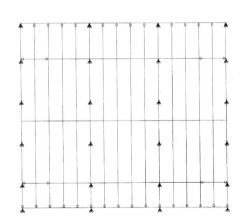
Plan

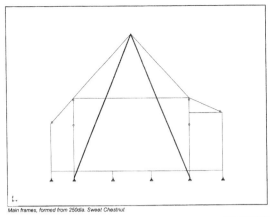
Main frames, formed from 250dia. Sweet Chestnut

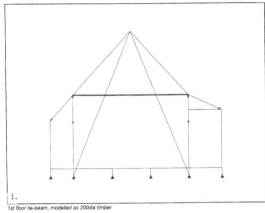
1st floor tie-beam, modelled as 200dia timber

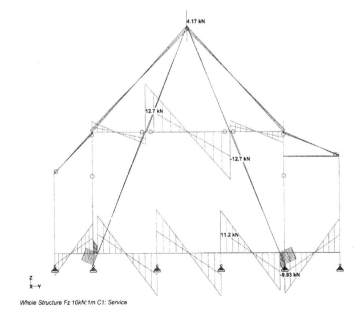
Whole Structure Fz 10kN:1m C1: Service

FIRST FLOOR JOISTS — MAX SPAN 3.0M

DESIGN MOMENT :- $(0.52+2.5) \times 0.4 \times 3.0^2 \times 0.125 = \underline{1.36 \text{ kNm}}$

LIMIT DEFLECTION TO $L \times 0.003$.

$I_{REQD} = \dfrac{5 \times (0.52+2.5) \times 0.4 \times 3.0^4 \times 10^2}{384 \times 11,000 \times 3000 \times 0.003 \times 10^4} = \underline{1287 \text{ cm}^4}$

$b = 47 \text{ mm}$

$d = \sqrt[3]{\dfrac{1287 \times 12}{4.7}} = \underline{14.87 \text{ cm}}$

TRY $50 \times 175 \text{ mm}$ dp joists @ 400%

$f_{bc} = (1.36 \times 6 \times 10^6) / (47 \times 172^2) = \underline{5.37 \text{ N/mm}^2}$

TIMBER — DOUGLAS FIR (HOME GROWN)
GRADE — (SS)
$p_{bc} = 6.2 \text{ N/mm}^2$
$K_2 = 1.0$
$K_3 = 1.0$
$K_8 = 1.1$
$K_7 \Rightarrow (300/175)^{0.11} = 1.06$

PERMISSIBLE BENDING STRESS :-
$\Rightarrow 6.2 \times 1.0 \times 1.0 \times 1.1 \times 1.06 = \underline{7.2 \text{ N/mm}^2} > 5.37$ OK

USE $50 \times 175 \text{ mm}$ Dp JOISTS @ 400%

GROUND FLOOR JOISTS — MAX SPAN 3000 mm

TIMBER — DOUGLAS FIR (HOME GROWN - SS GRADE)

DESIGN MOMENT :- $(4.0+0.3) \times 0.4 \times 3.0^2 \times 0.125 = \underline{1.94 \text{ kNm}}$

$p_{bc} = 6.2 \text{ N/mm}^2$
$E_{mean} = 11,000 \text{ N/mm}^2$
$K_2 = 0.8$
$K_3 = 1.25$ (MEDIUM TERM LL)
$K_8 = 1.1$
$K_7 = (300/197)^{0.11} = 1.047$

PERMISSIBLE BENDING STRESS :-
$\Rightarrow 6.2 \times 0.8 \times 1.25 \times 1.1 \times 1.047 = \underline{7.14 \text{ N/mm}^2}$

$Z_{REQD} \Rightarrow (1.94 \times 10^6)/(7.14 \times 10^3) = 271 \text{ cm}^3$

$b = 47 \text{ mm}$

$d = \sqrt{271 \times 6 / 4.7} = \underline{18.6 \text{ cm}}$

$I_{REQD} = \dfrac{5 \times (4.0+0.3) \times 0.4 \times 3.0^4 \times 360 \times 10^2}{384 \times 11,000 \times 0.8 \times 3000 \times 10^4} = \underline{2473 \text{ cm}^4}$

$b = 47 \text{ mm}$

$d \Rightarrow \sqrt[3]{\dfrac{2473 \times 12}{4.7}} = \underline{18.5 \text{ cm}}$

PROVIDE $50 \times 200 \text{ mm}$ dp JOISTS @ 400%

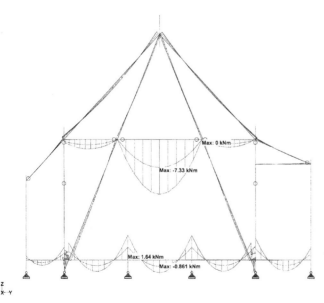

Whole Structure My 5kNm:1m C1: Service

FROM OUTPUT
MAX FORCE IN FIRST FLOOR BM - ELEMENT [29]
BM = -7.33 KNM ⎫
AXIAL = 6.06 KN ⎬ REFER TO CALC. 3.11A.
SHEAR = 12.7 KN ⎭

250 mm ⌀ SWEET CHESTNUT
$A = 491 \text{ cm}^2$ $p_{bc} = 7.6 \text{ N/mm}^2$
$I_{NA} = 19.2 \times 10^3 \text{ cm}^4$ $Z = 1534 \text{ cm}^3$ $q = 2.0 \text{ N/mm}^2$
                                           $p_c = 4.5 \text{ N/mm}^2$

$p_{bc} \Rightarrow 7.33 \times 10^6 / 1534 \times 10^3 = 4.8 \text{ N/mm}^2$
$f_q \Rightarrow (12.7 \times 10^3)/(491 \times 10^2) = 0.26 \text{ N/mm}^2$
$f_t \Rightarrow (6.06 \times 10^3)/(491 \times 10^2) = 0.12 \text{ N/mm}^2$

MODIFICATION FACTORS
$K_2 \Rightarrow$ (SERVICE CLASS 2) = NO REDUCTION REQD.
$K_3 \Rightarrow$ (LONG TERM) = 1.0
$K_8 \Rightarrow 1.0$
$K_6 \Rightarrow$ (SOLID CIRCULAR SECTION) = 1.18 (BENDING ONLY)

$p_{bc} \Rightarrow 7.6 \times 1.0 \times 1.0 \times 1.18 = 8.97 \text{ N/mm}^2$ Ratio = 0.54
$q \Rightarrow 2.0 \times 1.0 \times 1.0 = 2.0 \text{ N/mm}^2$ = 0.13
$p_c \Rightarrow 4.5 \times 1.0 \times 1.0 \times 1.0 = 4.5 \text{ N/mm}^2$ = 0.03
                                                     Σ = 0.7

NOTE: BM CHECKED AS SIMPLY SUPPORTED BETWEEN JOINTS HOUSING THE BEAM TO 'A' LEGS.

'A' FRAME
DESIGN LOADS - FROM OUTPUT.
MEMBER [97].
AXIAL LOAD = 23.3 KN
MOMENT ≃ 0 (NEGLIGIBLE)
SHEAR ≃ 0 (NEGLIGIBLE)

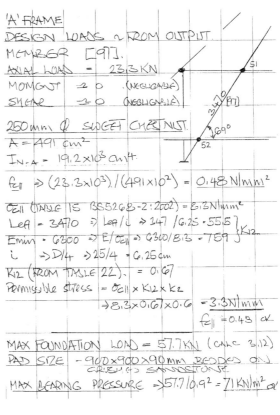

250 mm ⌀ SWEET CHESTNUT.
$A = 491 \text{ cm}^2$
$I_{NA} = 19.2 \times 10^3 \text{ cm}^4$

$f_{c\parallel} \Rightarrow (23.3 \times 10^3)/(491 \times 10^2) = 0.48 \text{ N/mm}^2$

$\sigma_{c\parallel}$ (TABLE 15 BS5268-2:2002) = 8.3 N/mm²
$L_{eA} = 3470 \Rightarrow L_{eA}/i \Rightarrow 347/6.25 = 55.5$ ⎫
$E_{min} = 6300 \Rightarrow E/\sigma_{c\parallel} = 6300/8.3 = 759$ ⎬ K_{12}
$i \Rightarrow D/4 \Rightarrow 25/4 = 6.25 \text{ cm}$ ⎭
K_{12} (FROM TABLE 22) = 0.67
PERMISSIBLE STRESS = $\sigma_{c\parallel} \times K_{12} \times K_2$
                  $\Rightarrow 8.3 \times 0.67 \times 0.6 = 3.3 \text{ N/mm}^2$
                                          $f_{c\parallel} = 0.48$ ✓

MAX FOUNDATION LOAD = 57.7 KN (CALC 3.12)
PAD SIZE - 900 × 900 × 90 mm BEDDED ON CRUSHED SANDSTONE
MAX BEARING PRESSURE $\Rightarrow 57.7/0.9^2 = 71 \text{ KN/m}^2$ ✓

CONNECTION - FIRST FLOOR BEAM / FRAME

VIEW A

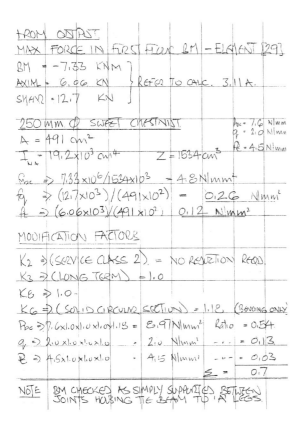

250mm DIA CRUX FRAME (SWEET CHESTNUT)
250mm DIA FLOOR BM. (SWEET CHESTNUT)
FLOOR BEAM SUPPORTS FIRST FLOOR JOISTS
CONTACT SURFACE
NOTE:- LEGS OF 'A' FRAME INCLINED @ 69°

PLAN

250MM. DIA. FLOOR BEAM
CONTACT AREA - A
[SHOWN WITH LEGS VERTICAL]
[ACTUAL AREA ELONGATED]
$r = 125 \text{ mm}$
$L \Rightarrow 125 - 45 = 80 \text{ mm}$
CHORD
$2\sqrt{(125^2 - 80^2)} = 192 \text{ mm}$
$\alpha = \cos^{-1} 80/125 = 50.2°$

AREA = $\left[\frac{100.4}{360} \times \frac{\pi \times 250^2}{4}\right] - (2 \times 80 \times 96 \times 1/2) = 6434 \text{ mm}^2$

CONNECTION [CONTINUED]

CONNECTION LOAD :-
$\frac{(0.52 + 3.0) \times 5.0}{4} \times 3.0 \times 1.15 = 15.18 \text{ KN}$ abitu

LOAD TRANSFERRED THROUGH DIRECT BEARING OF NOTCHED FLOOR BEAM ONTO RAKING LEG OF 'A' FRAME

FLOOR BEAM - 250 mm ⌀ SWEET CHESTNUT

COMPR. PERPENDICULAR TO GRAIN = 2.3 N/mm² ⎫ TABLE
COMPR. PARALLEL TO GRAIN = 8.3 N/mm² ⎬ 15

MODIFICATION FACTORS
$K_2 = 1.0$ (SERVICE CLASS 2)
$K_3 = 1.0$ (LONG TERM)
$K_6 = 1.10$ (FORM FACTOR - BENDING ONLY)

PERMISSIBLE COMPR. STRESS $\Rightarrow 2.3 \times 1.0 \times 1.0 \times 1.0 = 2.3 \text{ N/mm}^2$
CONTACT AREA = 6434 mm²
BEARING STRESS $\Rightarrow 15.18 \times 10^3 / 6434 = 2.36 \text{ N/mm}^2$
                                                    OK

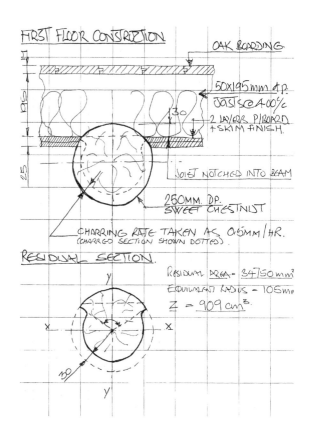

FIRST FLOOR CONSTRUCTION
- OAK BOARDING
- 50×195mm d.p. JOISTS @ 400 c/c
- 2 LAYERS P/BOARD + SKIM FINISH.
- JOIST NOTCHED INTO BEAM
- 250MM. DP. SWEET CHESTNUT
- CHARRING RATE TAKEN AS 0.5MM/HR. (CHARRED SECTION SHOWN DOTTED)

RESIDUAL SECTION.
- RESIDUAL AREA = 34750 mm²
- EQUIVALENT RADIUS = 105mm
- $Z = 909 \text{ cm}^3$

BENDING STRESS = $7.33 \times 10^6 / 909 \times 10^3 = 8.1 \text{ N/mm}^2$

MODIFICATION FACTORS.
$K_2 = 1.0$
$K_3 = 1.5$ (SHORT TERM)
$K_8 = 1.0$
$K_6 = 1.18$ (CIRCULAR SECTION).

$q = 2.7 \times 10^3 / 34750 = 0.37 \text{ N/m}$
$f_t = 6.06 \times 10^3 / 34750 = 0.17 \text{ N/m}$

		Stress ratio
$P_{bz} \Rightarrow 7.6 \times 1.0 \times 1.5 \times 1.0 \times 1.18 = 13.5 \text{ N/mm}^2$		0.6
$q \Rightarrow 2.0 \times 1.0 \times 1.5 \times 1.0 \times 1.0 = 3.0 \text{ N/mm}^2$		0.123
$P_t \Rightarrow 4.5 \times 1.0 \times 1.5 \times 1.0 \times 1.0 = 6.75 \text{ N/mm}^2$		0.025
	Σ stress ratio	0.75

RESIDUAL SECTION ADEQUATE TO SAFELY SUSTAIN FIRST FLOOR CONSTRUCTION FOR 1HR. FIRE DURATION.

APPENDIX B

ACKNOWLEDGEMENTS

Roundwood timber framing has been a journey for the past 10 years. From the seeds of early structures, the emergence of my Woodland House, through to the advancement of techniques in jointing and insulation by which Lodsworth Larder stands out.

I would like to thank the journeymen who have joined me on this quest. From their woodsman's lifestyle apprenticeship, many are now working for the Roundwood Timber Framing Company or creating roundwood buildings of their own. In particular Dylan, Rudi, Rich and Dave (the core team), Thomas, Barney, Andy (breaking through), Nick, Kris (raring to go). Chris, for his fine shingles, and many others who have turned out to cut a joint or lift a cruck and be a part of this new vernacular of natural architecture. To Langhams 'Hip Hop' for keeping the team lubricated throughout the builds.

I would like to thank all the team at Permanent Publications for their constant support. I would like to thank Kim Bold for typing up my original draft, Tristan Bund for his line drawings and Nicola Willmot for the excellent design of the book.

To my family for their unconditional support, my brother Dan and my dog 'Oily', for ensuring my hunting 'gene' is constantly active and my three children Rowan, Zed and Tess for keeping a smile upon my face.

A memory of Whitstable Bay, almost lost in the fast progression of life, like a seed awaiting its break from dormancy, now rekindled and flowing through my being.

APPENDIX C

GLOSSARY

Adze: An axe-like tool with its blade at right angles to its handle, used to shape and dress timbers.

Bender: Temporary home or shelter made from hazel branches and canvas covering.

Box Frame: A term used to describe a form of construction where the building is framed out of horizontal and vertical timbers to produce a wooden box.

Butterpat Joint: Scribed joint which resembles a pat of butter in a dish, where cruck meets the tie-beam in a roundwood timber frame, also referred to as dovetailed notch joint.

Cant Hook: Forestry hand tool based on leverage, enables the movement of large timbers.

Cladding: The covering that is applied to the outside surface of a timber frame.

Cleave: To split unsawn timber by forcing the fibres apart along its length.

Cogged Cruck Joint: The joint that secures the cruck poles together forming a pair of horns for the ridge pole to nestle in.

Coppice: Broadleaf trees cut during the dormant season, which produce continual multi-stems that are harvested for wood products.

Corner Chisel: A heavy duty L-shaped chisel struck with a mallet. Used for cutting and cleaning out corners of a mortise.

Cruck: Primitive truss formed by two main timbers, usually curved, set up as an arch or inverted V.

Disco Jowl: Also referred to as cranked jowl. A jowl post that has to bend around an underfloor support beam to pick up the wall plate in the right position.

Drawknife: A knife blade with handles on both ends so that the knife can be pulled by both hands towards the user.

Earth Anchors: Steel cable and anchor that are secured in the subsoil and then attached to the building to anchor it in place.

Feltch: To moisten a stiff fabric and then mould it to the desired shape.

Gable: The triangular vertical surface which continues the end wall of a building to the end of the ridge in a gabled roof.

Greenwood: Freshly cut wood.

Gouge: A curved chisel useful for following the curve on scribe lines on roundwood timber frame joints.

Hybrid Cogged Dovetail Joint: Joint between underfloor tie beam and cruck, named 'hybrid' as it is a joint where roundwood meets sawn wood.

Joist: Parallel timbers that make up the floor frame onto which floor boards are attached.

Joist Hanger: A metal fixing bracket which allows joists to be attached to a beam or attatched at 90 degrees to one another.

Jowl: The enlarged head of a post. (In this book the term is used for posts that take the traditional position of a jowl post in a historic timber frame).

Lath: Thin narrow strips of wood, usually cleaved from roundwood and used with plaster in walls and ceilings.

Mortise: A chiselled slot into or through which a tenon is inserted.

Permaculture: Ecological design for a sustainable future.

Ping: The process when a chalked string is stretched and manually released to mark a length of timber.

Ridgepole: A horizontal timber at the peak of the roof to which the rafters are attached.

Rafter: Inclined timbers following the slope of the roof, to which batons and shingles can be attached.

Roundwood: Timber that has not been sawn.

Scarf Joint: A joint for splicing together two timbers, end to end.

Scribing: To mark a timber by scratching or drawing a line and also to shape a timber so that it fits the irregular surface of another timber.

Shingles: Wooden tiles.

Slick: A long handled chisel, which is used like a plane to smooth off a surface.

Staddle Posts: Short posts that run from a padstone to support an underfloor support beam or underfloor tie-beam at intervals.

Standard: A single stemmed tree allowed to grow mature, commonly amongst coppice.

Stud: Lightweight timbers running vertically to help divide a wall. Insulation usually runs between them.

Sway Brace: In roundwood timber framing this is the term used to describe bracing between frames in the roof.

Tenon: The projecting end of a timber that is inserted into a mortise.

Tie-beam: A beam that spans the width of a building from wall plate to wall plate.

Underfloor tie-beam: A tie-beam that runs under the floor.

Underfloor Support Beam: A beam that runs the length of the building and supports floor joists and floor.

Wall Plate: A timber that runs horizontally along the top of the wall onto which the rafters are attached.

Wang: Term used to describe the flexibility and bend in a long round pole, often a wall plate or underfloor support beam.

Wattle and Daub: Branches, often hazel, woven around staves (wattle) and then covered (daubed) with a clay based earth plaster.

Wind Brace: In roundwood timber framing this is the term used to describe bracing between jowl posts and wall plates. (In traditional timber framing a wind brace is often the brace between a purlin and principal rafter.)

Yurt: A wooden framed transportable dwelling with canvas or skin covering originating from Asia, now often used as a dwelling in woodland.

APPENDIX D

FURTHER READING

Builders of the Pacific Coast; Lloyd Kahn; Shelter Publications, 2008

Building the Timber Frame House; Ted Benson; Fireside, 1995

Build It With Bales; MA Myhrman and SO Macdonald; Chelsea Green, 1997

Building With Logs; B Allan Mackie; Firefly Books, 1997

Building Your Own Low-Cost Log Home; Roger Hard; Storey Publishing, 1985

Digest 44; Building Research Establishment; CRC Ltd., 2002

Discovering Timber Framed Buildings; Richard Harris; Shire Publications, 1978

Energy Efficient Building: The Best of Fine Homebuilding; Taunton Press, 1999

Farm Buildings of the Weald 1450-1750; David and Barbara Martin; Heritage (Oxbow Books), 2006

Homework; Lloyd Kahn; Shelter Publications, 2004

Lime in Building; Jane Schofield; Black Dog Press, 1994

Low-Cost Pole Construction; Ralph Wolfe; Storey Publishing, 1980

The Natural Plaster Book; CR Guelberth and D Chiras; New Society, 2003

Notches Of All Kinds; B Allan Mackie; Firefly Books, 1998

A Pattern Language; Christopher Alexander, Sara Ishikawa and Silverstein Murray; Oxford University Press, 1997

Places of the Soul; Christopher Day; Harper Collins, 1990

Round Small-Diameter Timber For Construction; ed. Alpo Ranta-Maunus; VTT Technical Research Centre of Finland, 1999

Shelter; Lloyd Kahn; Shelter Publications, 1973

Timber Building in Britain; RW Brunskill; Gollancz, 1990

Timber Frame Construction; Jack Sobon and Roger Schroeder; Storey Publishing, 1984

Timber Pole Construction; Lionel Jayanetti; Intermediate Technology Publications, 1990

The Woodland Way; Ben Law; Permanent Publications, 2001

The Woodland House; Ben Law; Permanent Publications, 2005

The Woodland Year; Ben Law; Permanent Publications, 2008

APPENDIX E

RESOURCES

Adams Consulting Engineers Ltd,
177B Queens Road, Hastings, East Sussex
TN34 1RN
01424 442 399
www.adamsconsultingengineers.co.uk
infohastings@adamsconsultingengineers.co.uk
Structural engineers who have experience working with roundwood timber frames.

Anchor Systems International Ltd
01342 719 362
www.anchorsystems.co.uk
info@anchorsystems.co.uk
Suppliers of earth anchors.

Ashem Crafts
01905 640 070
www.ashemcrafts.com
enquiries@ashemcrafts.com
Producers of rounding and hollow shoulder planes.

Ashley Iles
01790 763 372
www.ashleyiles.co.uk
sales@ashleyiles.co.uk
Maker of woodworking tools, takes commissions.

Axminster Power Tools
0800 371 822
www.axminster.co.uk
cs@axminster.co.uk
Suppliers of many tools used in roundwood timber framing including scribers.

Ben Law
Prickly Nut Wood, Lodsworth, West Sussex
GU28 9DR
ben@ben-law.co.uk
www.ben-law.co.uk
The author runs short courses including roundwood timber framing.

Chapter 7
http://tlio.org.uk/chapter7
Chapter7@tlio.org.uk
Help, guidance and campaign for change in the planning system, for people genuinely working on the land or needing access to it. They also produce a very good magazine called *The Land*.

Green Shopping Catalogue
01730 823 311
www.green-shopping.co.uk
sales@green-shopping.co.uk
Books on ecological building, green woodworking, permaculture and many other related subjects. Also supplies quality tools and products.

Ray Iles
01507 525 697
www.oldtoolstore.com
ray@oldtools.idps.co.uk
Second hand green woodworking tools.

Timber Research And Development Agency TRADA
01494 569 600
www.trada.co.uk
membership@trada.co.uk
Organisation researching timber and its uses for the construction industry.

Windy Smithy
07866 241 783
www.windysmithy.co.uk
windysmithy@gmail.com
Maker of timber dogs and other timber framing tools.

Woodland Heritage
01428 652 159
www.woodlandheritage.org.uk
enquiries@woodlandheritage.org.uk
Organisation dedicated to improving the quality of trees grown in the UK. Holds study days, gives bursaries for forestry students and publishes an excellent journal.

Also by Ben Law

WOODLAND CRAFT
Accompany woodsman Ben Law as he celebrates the amazing diversity of craft products made from materials sourced directly from the woods.
From brooms, rakes, and spoons to chairs, fencing, and yurts, the items are hewn from freshly cut green wood, shaped by hand and infused with a simple, rustic beauty.
There are detailed instructions and advice for each craft, along with essential knowledge about tools and devices plus fascinating information on the history, language and traditions of the crafts, coppice management and tree species. Permanent Publications and GMC Publications. 216pp. Hbk. £25.00

THE WOODLAND HOUSE
Full of stunning colour photos, this is a visual guide to how Ben built his outstandingly beautiful home in the woods. It is also a practical manual and the story of a man realising a lifetime's dream to build one of the most sustainable and beautiful homes in Britain. Permanent Publications. 96pp. Pbk. £16.95

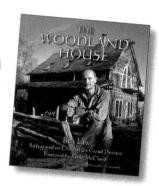

THE WOODLAND YEAR

Ben is arguably Britain's greatest living woodsman... He is a true pioneer and is quite simply creating a woodland renaissance in Britain. Read this, and you will surely want to be part of it.
Hugh Fearnley-Whittingstall, from the book's foreword

Packed with stunning colour photographs, *The Woodland Year* provides a fascinating month by month insight into every aspect of sustainable woodland management; the cycles of nature, seasonal tasks, wild food gathering, wine making, mouth watering and useful recipes, coppice crafts, roundwood timber frame building (pioneered by Ben in the UK), nature conservation, species diversity, tree profiles and the use of horses for woodland work. Each month also includes guest contributions from woodlanders in other parts of England and Wales.

This is a profound book that is both practical and poetic. It describes a way of life that is economically and ecologically viable and sets a new standard for managing our woods in a low impact, sustainable way. As such, it holds some of the fundamental keys to how we can achieve a lower carbon society. Permanent Publications. 176pp. Hbk. £24.95

THE WOODLAND WAY
A Permaculture Approach To Sustainable Woodland Management

For everyone who loves trees and woodlands. This radical yet practical book presents an immensely practical alternative to conventional woodland management. It clearly demonstrates how you can create biodiverse, healthy environments, yield a great variety of value added products, provide secure livelihoods for woodland workers and farmers, and benefit the local community. Permanent Publications. 256pp. Pbk. £24.95

Enjoyed this book?
You might also like these from Permanent Publications

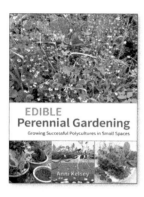  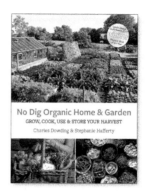

Edible Perennial
Gardening
Anni Kelsey
£15.95
Explains how to source
and propagate different
vegetables, which plants
work well together in a
polycultures.

The Creative Kitchen
Stephanie Hafferty
£19.95
How to make seasonal,
plant-based meals,
drinks, soaps, balms,
and store cupboard
ingredients like vinegars
and essences.

No Dig Organic
Home & Garden
Charles Dowding and
Stephanie Hafferty
£23.00
Award winning guide to
creating and growing in a
no dig garden, including
harvesting, preserving
and cooking your produce.

Our titles cover:

permaculture, home and garden, green building,
food and drink, sustainable technology,
woodlands, community, wellbeing and so much more

Available from all good bookshops and online
retailers, including the publisher's online shop:

https://shop.permaculture.co.uk

with 10% off the RRP on all books

Our books are also available via our American distributor, Chelsea Green:
www.chelseagreen.com/publisher/permanent-publications

Permanent Publications also publishes *Permaculture Magazine*

Enjoyed this book? Why not subscribe to our magazine

Available as print and digital subscriptions, all with FREE digital access to our complete 26 years of back issues, plus bonus content

Each issue of *Permaculture Magazine* is hand crafted, sharing practical, innovative solutions, money saving ideas and global perspectives from a grassroots movement in over 170 countries

To subscribe visit:

www.permaculture.co.uk

or call 01730 776 582 (+44 1730 776 582)